The
Amazing
World
of
ANIMALS

contributions by

Dr Michael Stoddart
Richard Martin
Robert Burton
Dr Ulrich Gruber
Alwyne C. Wheeler
Michael Chinery
Dr Roger Hamilton

edited with a foreword by

Sir Peter Scott

The Amazing World of ANIMALS

NELSON

Thomas Nelson and Sons Ltd
Lincoln Way, Windmill Road,
Sunbury-on-Thames, Middlesex
P.O. Box 18123, Nairobi, Kenya

Thomas Nelson (Australia) Ltd
19-39 Jeffcott Street, West Melbourne 3003

Thomas Nelson and Sons (Canada) Ltd
81 Curlew Drive, Don Mills, Ontario

Thomas Nelson (Nigeria) Ltd
P.O. Box 336, Apapa, Lagos

First published in Great Britain by
Thomas Nelson and Sons Ltd in 1976

Concept, picture research and picture
editing: Alexander Low
Line drawings and maps: Ron Hayward

Produced by London Editions Ltd,
30 Uxbridge Road, London W12 8ND,
England

ISBN 0 17 149046 0

Printed and bound in Spain by TONSA - San Sebastian

Dep. Legal: S.S. 34 - 1976

Contents

Foreword by Sir Peter Scott 7

Mammals 13

  Introduction to mammals by Dr Michael Stoddart 13

Large mammals by Richard Martin 17

  The great apes 17

  Bears 20

  The giant panda 21

  Tigers and lions 22

  Elephants 24

  Whales 25

  Rhinoceroses 27

  Camels 28

  The giraffe 30

  Other ruminants 31

Smaller mammals by Dr Michael Stoddart 40

  The beaver 41

  Mammalian reproduction 43

  Desert mammals 45

  Rodent population cycles 46

  Sub-culture in Japanese monkeys 48

  Anteaters 48

  Marsupial 'mice' 49

  Prairie dogs 50

  Bats 51

  The honey badger 52

  Dogs and foxes 53

  Rodents 55

Birds by Robert Burton 64

  Introduction to birds 65

  The lengthy courtship of albatrosses 67

  Feeding helps togetherness 68

  Competition for females 69

  Where to nest 70

  Weaving nests 71

  Communal families 72

  Mimicking for survival 73

  Voice keeps the family together 74

  Sharing out the food 75

  Oystercatchers learn to feed 76

  Stone-throwing vultures 77

  Butcher birds 78

  Finding the way 78

  Fuel for migration 79

Contents

Amphibians and reptiles by Dr Ulrich Gruber   **88**

Introduction to amphibians and reptiles   89
Salamanders that always remain as larvae   93
Spadefoot toads   94
Frogs with eggs on their backs   95
Some strange forms of turtle   96
Crocodiles   97
The tuatara   98
Chameleons   100
Monitors   101
Gaboon viper and tropical rattlesnake   102
Snakes that climb trees   103

Fishes by Alwyne C. Wheeler   **112**

Introduction to fishes   113
Strange associations   115
Fish migrations   116
A four-eyed fish and other fishes' eyes   117
The unsilent sea   118
Colours and patterns   119
Spears, stings and shocks   121
Mimicry and protective resemblances   123
Cleaner fishes   125
Fish-life in the deep sea   126

Insects by Michael Chinery   **136**

Introduction to insects   137
Insect behaviour   141
Honey bee language   142
Farmers and herdsmen   144
Slave-making ants   145
Weaver ants   146
Trap-building insects   147
The mating game   147
Warning colours and mimicry   149
False eyes and heads   150
Jamming the bat's radar   151
Strange partnerships   151

Evolution by Dr Roger Hamilton   **161**

Mammals   162
Birds   164
Fishes   166
Amphibians   168
Reptiles   170
Insects   172

Photographic credits   174
Index   175

Foreword

by Sir Peter Scott

Life on earth, so the palaeontologists tell us, began at least 2000 million years ago – twenty million centuries, and perhaps as much as twenty-seven million centuries ago. A time-scale of that order is difficult to comprehend, but we and all other living things on earth today are the end-products of an evolutionary process which has been going on for that length of time. During those aeons the original life forms have gradually, by an endless series of changes, some small, some large, given rise to the almost incredible diversity of plants and animals which now share the planet with us.

As a painter and a naturalist I have throughout my life derived enormous pleasure from observing that diversity, not only of form and colour but of life-style and behaviour. Very often I have found myself overcome with sheer wonder at the unimaginable beauty and intricacy and ingenuity of nature. The huge gaps in the fossil record make it impossible to conceive all the evolutionary steps which have led in a continuous sequence of mutations to the amazing perfection that is now there for us to see, and I am left with a sense of deep reverence for the exquisite process of evolution by natural selection – a feeling which sometimes amounts almost to a mystical experience.

Consider how it has come about that one species of bush-cricket in Central America so perfectly mimics a leaf as to have evolved on its wing a small dark patch exactly resembling a patch of mould, while a related species has on its wing a perfect reproduction of the appearance of a leaf whose surface has been partially devoured by newly hatched caterpillars. At what precise point did one of the ancestral bush-crickets set off on the 'mould' mimicry and another on the 'caterpillar-eat' mimicry? The protection they each achieve must have some relationship to their overall numbers, for if too many look uniformly mouldy the birds or reptiles will learn that leaves with that kind of blemish are good to eat.

Consider the hawkmoth caterpillars which, after their third skin moult, may emerge in one of two or more colours, with corresponding differences in behaviour. This applies to a number of species in most of which the caterpillars are either basically green or basically brown. Each colouring gives protection from certain predators: green from birds and brown from lizards. At one time of year the birds are more numerous, at another the lizards, and the proportions of green to brown in succeeding broods apparently vary accordingly. The statistical balance has been computed by nature with exquisite accuracy. If it were not so the hawkmoths would no longer be with us.

Consider the orchids which need to be pollinated by insects. Several of them have evolved flowers which are not only like the females of the insects concerned, but are even more attractive – a super-stimulus to the males; and think of the time scales involved to achieve those changes. No wonder we have to think in millions of years.

And consider also the animals and plants that reached a high degree of perfection far back in time and have needed to change little, if at all, to survive ever since. The amoeba still soldiers on with its single cell, and so do a host of unicellular animals and plants that must have looked almost the same at least 1500 million years ago. Fossil amber has preserved an ant from the Eocene, not less than 33 million years old, which has on its antennae precisely the same adaptation for milking aphids as a living ant of today, and appears indeed to be the same species. Great numbers of species have long ago reached a stage of such perfect adaptation to their particular mode of life that little if any further mutation has been necessary for their successful survival.

The title of this book invites us to look at the world of animals and be amazed. I have been looking at that world for sixty years and remain constantly amazed. Man probably has a more highly developed sense of curiosity than any other animal yet every question that we answer about animals poses half a dozen more, and for the most part the answers are unexpected and continually amazing. Most often the source of amazement is how such an intricate pattern of form or colour or behaviour could have been reached.

It is perhaps this amazing perfection wrought by natural selection over the millions of centuries which, most of all, has brought me to a deep concern about species extinction. Of course species have been becoming extinct since the dawn of life and far more of them have disappeared than are extant today. The fossil record is largely the scrap-heap of unsuccessful end products – evolutionary production lines that failed. The profusion of great reptiles which lived in Jurassic and Cretaceous times – the age of dinosaurs – may well have been brought to an end by our ancestors, the emergent mammals. But now one species, man, has so changed the surface of the planet as to increase the 'normal' rate of extinction by what has been estimated as a factor of between twelve and sixteen. In the last 400 years twelve to sixteen times more species have been exterminated at the hands of man than would have died out if modern man had never existed.

If modern man himself could not survive without such extermination of the fellow crew-members of his spaceship, there might be some justification for this destructiveness. But it is not so. In all but a very few cases the species have become extinct through human ignorance, greed, or apathy. What harm did the dodo or the great auk, or Steller's sea cow ever do to mankind? Who was ever bothered by the gentle Labrador duck or the blue antelope? Had the people who killed out the last of those animals known what they were doing, and been aware of the uniqueness of the species and the total irrevocability of extinction, they might have desisted, although at that time no ethic of conservation had yet emerged. The evolution of human behaviour had not yet reached that stage, and even now it has been accepted by only a very small proportion of the world's human population. There is a long way to go before these ideas are universally acknowledged by all the world's people. And there is another hurdle to jump, for the extinct animals I have mentioned were harmless to man's interests, and indeed supplied him with food. What about the animal species which are positively harmful – those which kill and eat people or their domestic animals, those which compete directly with man for his food supplies, those which are vectors of his diseases? Ethically the conclusion must surely be the same. Thou shalt not exterminate. At the material level the argument against extermination can only be maintained on the ecological plane: that the web of life is complex, that the food chains are the time-honoured product of evolution, and that removing, for example, the large predators, or all insects in order to destroy certain pest species, may not after all be in the long-term best interests of mankind as a whole. Alas, there are few human societies which are in any case concerned with the long-term best interests of mankind as a whole. And in the background there is the basic arrogance of humanity, which has long believed that it has sole rights to planet Earth, and has more recently tended to become convinced that it already 'knows it all'.

The evolution of human behaviour still has a long way to go before it reaches the perfection of suitability to its mode of life achieved by the Eocene ant. But the evolution of conservationists and environmentalists as a significant moral and political force in the last fifty years is perhaps an encouraging sign. The time scale of evolution is inevitably slow. It is the kind of information contained in the pages of this book which is especially needed to help speed it up. For the book deals with many up-to-the-minute discoveries about all four vertebrate Classes of animals – Mammals, Birds, Amphibians and Reptiles, and Fishes. It also deals with one, and much the biggest of the invertebrate Classes – the Insects. It is a sobering thought that although a million species of insects have been scientifically 'described' and named, there are estimated to be another million which have not. The entomological taxonomists have a lot of work ahead.

What this book tells cannot fail to arouse just that additional sense of wonder and amazement which is needed to bring new support to the principal ethic of conservation – that the other animals and plants have as much right to exist as we have. Even if (and perhaps because) we are the only living beings to know it, man surely has some sort of responsibility to safeguard these other marvels of the evolutionary process, at least against the carelessness and arrogance of his own species.

The great reptiles flourished in Jurassic and Cretaceous times, the mammals came to their climax – or would be coming if the very capacity of the biosphere to support life were not threatened by man. Birds have become adapted to life on almost all parts of the earth's surface. They can live in polar regions, far out in the oceans, in deserts and in caves, they can move on and under the water, but above all they can fly, and it is surely this capacity which especially delights and fascinates us. Since early youth I have found the migration of birds a source of wonder. I have followed the studies of bird navigation which are still yielding new information, though none, to my mind, more astounding than the evidence that the distribution of stars in the night sky provides the basis for orientation by night migrants and that the ability to use this immensely intricate pattern is conveyed to succeeding generations, as with all the other code messages of heredity, in a molecule of DNA (deoxyribonucleic acid). The young nightingale, or willow warbler, or blackcap must set off alone in the autumn, flying at night to avoid being caught by a hawk and flying towards Africa by the stars. Its parents may well still be feeding the second or third brood of the summer, and, in contrast to wild goose and wild swan families, they will in no case be leading their young on the migration flight.

But with all their mastery of the air, birds have never got away from the problems of laying and, with a few exceptions, incubating their eggs. For this they have to come down to earth, or at least to a nest in a tree. It is a limitation which the fishes in their element have to some extent overcome. The fishes – especially the bony fishes of the tropics – are also at some sort of climax of diversity in speciation. They have many of the characteristics of birds; indeed it is not too fanciful to think of them as underwater birds. In behaviour studies knowledge of the one illuminates the other.

It is in these two groups of animals, the birds and the fishes, that I have myself found the greatest interest and enjoyment. Waterfowl have been my life-long speciality – but it is less than twenty years since I first dived on a coral reef and my preoccupation with coral fishes began. That dive – or rather swim, (because most of the time was spent on the surface looking downwards, as it still usually is when I swim over coral) – was on the Great Barrier Reef, which is reputedly the most varied and colourful in the world. Since then I have swum on a great many coral reefs all over the tropics and have found some in Indonesia which are even more beautiful than the Great Barrier. What remains perennially exciting about swimming over coral reefs is the realization of just how little is known, just how few trained zoologists have ever seen what you are looking at. New observations can be made almost every day, which modify and sometimes contradict the accepted order of marine biology, for in that science the study of the living fish in its natural environment is still in its infancy. Among the wrasses and parrot fish we are still trying to discover which are females, which are males and which are super-males of the same kind, for in several species there is, it seems, only one brightly coloured male in each social group while there may be many other adult males in much the same colouration as the females. If this super-male is removed from the group another assumes the bright colours. The issue is further complicated by the capacity of some species to change sex. It has lately been discovered that a number of species including for example the little cleaner fish, *Labroides dimidiatus* or blue streak, are all born females and only the largest in the group turns into a male. If anything happens to him the next largest becomes a male, and so on. Elucidating these problems is not made easier by the capacity of many species of fish to change colour and pattern in a matter of seconds, and a further complication is the existence of a number of constant but widely different colour phases within the same species.

Nowhere, in my experience, is the sense of wonder at the world of animals more intense than when swimming over a coral reef. Yet on all too many of them man's irresponsible thoughtlessness is in evidence. Terrible depredation of the coral is taking place for the souvenir trade, and on most of the accessible reefs the larger fish have already been speared out. There is urgent need for underwater parks, reserves and sanctuaries, so that the marine habitats are not wantonly destroyed as so many above-water habitats have been.

We take it for granted that great buildings and sculpture and paintings by the Old Masters must be preserved. Many of them are hundreds of years old and irreplaceable. To destroy them would be inexcusable vandalism. Yet, strangely, species of plants and animals that are equally irreplaceable, and millions of years old, do not automatically evoke the same attitudes of reverence and responsibility as do works of art. Whatever theological views we may hold – whatever we may believe about nature's grand design and whether it postulates a Grand Designer – whether or not we regard the patterns of our planet and our universe as the creations of God, they exist; and in my philosophy it cannot be right to place the works of man above them. As a painter with almost infinite reverence for the artistic giants of human history, I must categorically repudiate such an inversion of values. The art galleries have their place, but it is the earth that needs our care and nurture, and a new awareness . . .

I believe this book will increase the awareness of its readers and instil in them the desperate urgency of safeguarding the amazing world of animals.

Slimbridge
May 1975

Introduction to
Mammals

by Dr Michael Stoddart

Mammals can be distinguished from all other animals by the facts that a) they all have hair, b) they feed their young on milk, and c) they have a heart which is divided into four chambers so allowing a high metabolic rate and an active life. A few species appear to be hairless, eg. whales and mole rats, but on close examination this is found to be untrue—juvenile whales have a small bristly moustache and mole rats have hairy tails. The particular form and characteristics of the various mammalian species is directly related to the ecological niche occupied by the species. Paradoxically the greatest single characteristic of the mammals is their variability, and the way they have filled almost all ecological niches. Mammals can be found in the air (bats), in the canopies of the highest trees (monkeys, squirrels), browsing on the forest floor (deer), grazing on the plains (antelopes, cattle), in high mountain places (sheep, goats, marmots), as herbivores (rodents) and insectivores (shrews) in the leaf litter on and under the soil surface, as active predators (dogs, cats, martens), as fresh water herbivores (dugongs) and salt water filter feeders (baleen whales), and as fresh and salt water fishers (seals, toothed whales). There are no strictly parasitic mammals, although vampire bats are close to being parasitic, and there are no day-flying insectivorous mammals to rival the birds. This amazing spread of niches is made all the more remarkable by the fact that there are only 4200 species of mammals, compared with 6000 species of reptiles, 10,000 birds, 20,000 bony fishes and probably well over 2 million insects. By any measure, the class of mammals is a highly successful one.

The size range of mammals is most dramatic. The smallest species, the Etruscan shrew, weighs $\frac{1}{25}$oz (1.5g); the largest, the blue whale, weighs up to 150 tons (153,000kg). Although this range is enormous it is not as large as that found in the insects which go from a few micrograms (0.000,000,072oz) to 1oz (30g). The largest land mammal, the African elephant, can weigh $4\frac{1}{2}$ tons (4570kg). It is doubtful if a larger mammal which supported its weight on four legs could ever exist on land. As it is, elephants are only able to amble along, and real running is out of the question. Mammals are warm blooded and the very large ones have a serious heat problem. The laws of physics dictate that the larger a body becomes, the smaller, relatively, its surface area becomes. As most excess heat is radiated from the skin, very large species are in constant danger of overheating. Thus they lack the hairy insulative coat of smaller species and they spend as much time in the shade as possible. Elephants' ears are little more than radiators which are flapped backwards and forwards in an attempt to bring about cooling. It is extremely doubtful if the large whales could exist in warm tropical seas—their vast bulk necessitates their being voluntary prisoners in the freezing polar oceans.

Depending on what type of food they eat, mammals' teeth are highly specialized. The basic number is 44, and that is found in shrews, hedgehogs, moles and their like. Ant-eating mammals, of which there are a great many (giant anteater, echidna, pangolin etc.) have no teeth at all—the insects are swept into the mouth by a long sticky tongue and swallowed whole. Blood-sucking vampire bats have four teeth—razor sharp canines—with which to perform their delicate surgery on their unfortunate hosts. Herbivores do not have a full set of teeth—usually the canines are lacking and a long gap, called the diastema, separates the incisors from the cheek teeth. Carnivores have little need of cutting teeth, so these tend to be rather small. Their eye teeth are huge, however, and are used for stabbing their prey. Development of these reached a peak in the now extinct sabre-toothed tigers which had canines up to 12in. long. Seals and whales which eat slippery fish have many teeth, all very sharp and pointing backwards. The record number of teeth for any mammal is held by the killer whale, which has 260!

Mammals catch their food in a variety of ways. Grazing and browsing species have little trouble since their food does not move. Plant material is not as nutritious as meat, so herbivores need to spend much time in feeding. Elephants, which consume up to 700lbs of leaves a day, spend about twenty-

Sizes of some mammals, compared to a blue whale (108ft long). *Left to right:* African elephant (11ft 6in. at shoulder); red fox (2ft at shoulder); man (5ft 8in. tall); lion (9ft long, including 2ft 6in. tail); red deer (5ft at shoulder); giraffe (18ft tall); porpoise (7ft long)

three out of twenty-four hours feeding. Most herbivores spend about six hours a day feeding. The cloven-hoofed species (antelopes, cattle) are at a distinct advantage over the single-hoofed species (horse), because nature has provided them with a multi-chambered stomach. A brief trip to the feeding ground allows one of these chambers to be filled with unchewed grass. Then, in the seclusion of a safe sheltered place, the cud can be regurgitated and chewed up properly. Horses must stand out in the exposed open plain and chew each mouthful forty-four times before swallowing it. It is thus not surprising that they are able to run much faster than cattle; speeds of up to 50mph have been recorded. Kangaroos, which are the Australian ecological equivalent of grazers, can run at 30mph and more. Quite by far the fastest speeds are achieved by the large carnivores, which utilize the tactic of ambush for hunting success. Cheetahs prey on small gazelles which are themselves capable of 30mph. In order to overhaul them, cheetahs produce bursts of up to 80mph, but this can be maintained for only a hundred yards or so. (The pronghorn, a type of antelope of North America can, however, maintain a somewhat slower speed for longer distances.)

Social life is of great importance to the mammals. Some of the carnivores live almost solitary lives, but most species live in family groups, herds or large populations. Half a century ago herds of bison in North America numbered $2\frac{1}{2}$ million and of springbok in South Africa 4 million. Early zoologists in South Africa reported that mass migrations frequently occurred

and ended in countless animals being drowned in the sea, so great was the pressure of numbers. A line of bodies on the shore, 6ft high and 30 miles long, was not a very unusual sight. Alas, both species have borne the brunt of the settlers' guns and such spectacles are gone for ever. Within their groups mammals spend much time engaged in social competition and contests of rivalry. These, like the rutting season clashes between stags, are ritualistic affairs involving a harmless clash of antlers. Blood is seldom drawn. Leadership of the herd and choice of mates are the prizes for the successful. Success may be short lived, for the constant pressure from usurpers is immense.

Of the 4200 species, almost half are rodents. The majority of these are small and weigh between 2 and 8oz (60-240g). Being mostly grass, grain and plant eaters they constitute a major threat to man's agronomy. Since most reproduce six times a year and have litters of up to ten young in each, their populations can double every few months. In spite of hundreds of years of fighting rats in Britain, and at an immense cost, there is still one rat alive for every man, woman and child. Rodents are so adaptable that they can accommodate to most forms of attempted control and poisons are only effective for a few years at the most. Recent research has shown that even the efforts of the wartime Manhattan project on Bikini Atoll in the South Pacific failed to exterminate native rats, for a strain resistant to the effects of radioactive fall-out evolved soon after the tests. Not all mammals are harmful to man and the potential of some as a source of protein is slowly being realized. The sales of bushmeat, naturally occurring species such as giant rats, green monkeys and bushbuck, in East Africa now runs at $160,000 (£66,000) a year and could be considerably increased. Naturally occurring species are much more resistant to disease than exotic, imported species. By interbreeding antelopes such as blesbok with cattle, offspring are produced which are resistant to disease like rinderpest. A few species, such as the eland, are tameable, and herds are now established in the USA and USSR. It is said their milk is particularly beneficial to people suffering from stomach and duodenal ulcers.

In short, mammals are a remarkable group of animals whose full biological and economic potential we are just beginning to appreciate. Alas, in the case of many species the scales have fallen from our eyes too late. Active conservation and the elimination of illegal hunting for fur and ivory will save some of the others. They will serve not just our aesthetic needs but, with careful management, our alimentary needs as well.

Large mammals

by Richard Martin

The great apes

The origins of man are the same as those of the great apes, but we must not assume that he descended from apes. It is generally accepted that man's nearest relative in the animal kingdom is the chimpanzee (*Pan troglodytes*), which inhabits the equatorial forests of Africa. This is certainly the most intelligent of non-human primates and is also known to use rudimentary tools, such as sticks and stones, either for defence or in acquiring food. Under controlled conditions chimpanzees are well able to reason out simple problems and they display a definite sense of organized design. However, they lack powers of concentration and soon become bored in most contrived situations.

Highly social, the chimpanzee lives in troops comprising many individuals of different sexes. It possesses a wide vocabulary of calls and signs and can be taught to communicate with humans by means of signs—though attempts at verbal communication have by and large failed. It spends much of its time at ground level foraging for vegetable material and insects, but recent research has shown that it is also capable of killing and eating animals up to the size of a pig or a monkey. Where, as in the rain forests, food is plentiful, chimpanzees will band together in huge communities and seldom bother about travelling far. In such circumstances, eight square miles is an average home range, though this may be increased to more than thirty square miles in the drier savannah.

The sexual behaviour of chimpanzees is particularly interesting since it appears to contrast with the social structure of the troop. In the troop, while it is true that there is no strict hierarchy, there are certain rules of dominance which all members respect. However, a female in season will be mated by a number of males, and there appears to be little or no sexual rivalry between them. Male dominance serves only a social function, and it is established by noisy and energetic displays.

Families of chimpanzees remain close-knit throughout life, youngsters remaining with their mothers for about six years, though some, especially adolescent males, will wander away and join other troops. This avoids the danger of in-breeding. Furthermore, it has been recorded that close relatives show a definite disinclination to mate with one another.

Chimpanzees are the most playful of apes and much has been recently learnt about their behaviour by the studies of Jane van Lawick-Goodall and others. Her absorbing and enlightening findings have given impetus to similar

studies of the gorilla (*Gorilla gorilla*)—the largest of all primates.

The Gorilla is massive, compared with the chimpanzee, and may weigh three times as much. Averaging 5ft 6in. to 6ft, when mature, it can weigh up to 500lbs. In common with its other anthropoid ape relatives, its arms are long giving a span of some 8ft. In appearance it is magnificent and terrifying. It was only discovered in 1861 and accounts of its ferocity were greatly exaggerated. Its strength and power are clearly apparent, but it is normally a gentle, peaceable animal and, unlike the chimpanzee, totally vegetarian. (In captivity, however, possibly through suffering the same type of deprivation found among human prisoners, it can become sullen and dangerous.)

The gorilla (centre) is by far the heaviest of the primates. A large male may weigh as much as 450lbs and stand 6ft, though females are generally only half the weight. The smallest of the apes is the gibbon (right), standing only 3ft in the common gibbon and as little as 15in. in the dwarf siamang of the Mentawei Islands, west of Sumatra. Gibbons have enormously long arms, with which they swing through the branches, sometimes in 40ft leaps. The chimpanzee (second from left) is the most familiar and studied of all the apes. Weighing up to 120lbs and standing about 4ft 6in., they display a social structure closer to man than any other. Orang-utans (second from right) live in Borneo and Sumatra. Despite protection they are rare creatures. Larger than chimpanzees, a male may weigh as much as 200lbs and stand 5ft 6in. They are covered with shaggy red hair and, like gibbons, enjoy a totally arboreal existence. Their legs are disproportionately short. Man himself is the least hairy of the primates. Standing an average 5ft 8in., his skeleton is modified for his upright position. His brain is far larger than that of other primates—1500cc in man; 510cc in the gorilla; 450cc in the orang-utan; 420cc in the chimpanzee

It spends ninety percent of its time near or at ground level, though the younger and more active members will retreat into trees. It fills its days in troops of from eight to thirty individuals, foraging over an area of perhaps fifteen square miles, only ascending trees to pluck fruit, rest and sleep.

Lowland gorillas are concentrated in the forests of tropical west Africa, between the Niger and Congo rivers. The much rarer mountain gorilla lives high in the densely forested mountains west of Lake Victoria, and on the Mitumba Mountains in tracts isolated by deforestation and agricultural development.

Gorillas probably have as tight a social structure as chimpanzees. Each troop is led by a silvery-backed male at least ten years old. This leader is respected by all the lesser members—none of whom would dare question his decisions or stand in his path. But despite his authority a dominant male gorilla is still a gentle animal. For all this strength (or perhaps because of it) fights are almost unheard of within a troop. If subjected to severe provocation a display of aggression—the much parodied rearing up and chest beating—is enough to frighten away any but the most foolhardy adversary. Humans are usually quite safe as long as they hold their ground and do not try to run away. The male gorilla may rush to within three feet, but will then invariably retire. But native hunters who return to their villages displaying gorilla bites on their posteriors are open to ridicule by their comrades for cowardice.

As with chimpanzees, the males show no sexual jealousy. Copulation is a leisurely affair which can take up to an hour before consummation. The gestation period is almost as long as that of humans, but the new-born gorilla baby weighs only a pound or two. The maternal instinct is strong and is manifested in the care and tenderness to which the baby is treated.

Though small when born, gorilla babies grow twice as fast as humans. They are very active by five months and by seven months can climb nimbly. They continue to suckle for nearly two years, but because of the slow reproductive rate—a female giving birth only every third or fourth year—there is time for the young gorilla to reach adolescence and a degree of independence before its parent's attention is diverted to another baby.

The only great ape to be found outside Africa is the orang-utan (*Pongo pygmaeus*), which lives in the sultry forests of Sabah, Borneo and Sumatra. The orang-utan (which means 'old man of the woods') is intermediate between the gorilla and the chimpanzee in size (about 4ft tall) and spends far more of its time in trees. It is a striking animal to see, with its shaggy covering of rust-red hair. Old males sport large cheek pouches and full beards and often take to a solitary existence, becoming unpredictable and dangerous.

Orang-utans swing slowly and deliberately through the branches and like their African relatives construct comfortable sleeping platforms at dusk— sometimes as high as 70ft above the ground. These hairy fruit-eaters are unfortunately becoming rare, and despite protection by some governments, the adults are often killed by poachers so that the babies can be taken and the young sold for considerable sums to zoos all over the world. However, international agreement between the major zoo bodies has helped to minimize this danger, and orang-utans are being bred in zoos with more and more success.

Bears

While chimpanzees, among the apes, are developing carnivorous or meat-eating tendencies, as we have seen, bears, though firmly placed within the animal order of Carnivora, are omnivorous—eating both meat and plants. They are unusual among carnivores in that they are very bulky and have a slow, shambling gait. Obviously to catch live animals for food, carnivores must either be small enough to skulk under cover and stalk their prey or fast enough to outrun it. The majority of carnivores (some 150 species) are therefore of the mongoose/weasel type, while others are scavengers or small cats and foxes.

Bears are fairly equally distributed between the Old and New Worlds, though they have strangely failed to colonize Africa. Most remarkable of all is the polar bear (*Thalarctos maritimus*), found round the desolate shores of the Arctic Ocean. It is also the largest bear, standing up to 5ft at the shoulder and reaching 9ft in length in the biggest specimens. More carnivorous than its relations, and totally dependent on the freezing sea for its food, it has adapted its basic bear characteristics to cope with the severity of its environment and to become an extremely capable swimmer. The most noticeable examples of this adaptation are its camouflaging white colouring, dense fur, small external ears and hairy soles to its feet, though these may well be more useful as anti-slip pads for walking across ice than for keeping warm.

The polar bear's favourite food is the little ringed seal (*Pusa hispida*), which it stalks over the ice and amongst the floes, or ambushes at its breathing hole. A single blow from the bear's powerful forearm crushes the seal's skull and kills it instantly. It is then said to pull the dead seal up through its breathing hole with such force that its pelvis may be crushed.

Before the Arctic winter has set in, the expectant polar bear mother will have excavated a den beneath a snow drift. Here she will spend most of the winter sleeping, conserving her energy and living off accumulated fat until her twin cubs are born in about December. They will remain in this snug, igloo-like den until spring, protected from the surrounding temperature (as low as 14°F) by the warmth generated by their mother's body.

Polar bear cubs are incredibly small when born, weighing only about 1½lbs, but they grow quickly on the mother's rich milk, and by the time spring arrives, three months later, they are well able to emerge into the outside world. Each cub will by then weigh as much as 25lbs, but the mother will have lost as much as 700lbs. Mating between the adults takes place soon after the sexes are re-united in the spring, but the fertilized egg does not become implanted in the wall of the uterus probably until September. By this means a midwinter birth is ensured, and this in turn guarantees that the whole of the

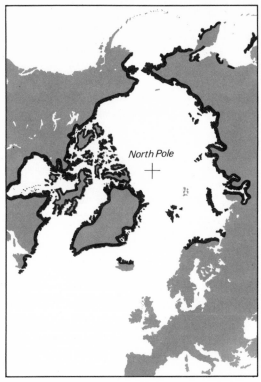

North Pole

Map showing the distribution of the polar bear around the Arctic circle (heavy lines)

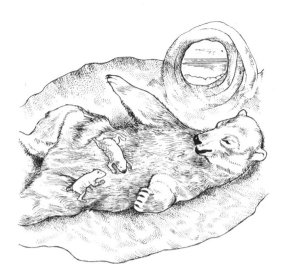

Polar bear cubs, usually twins, are born in mid-winter in a den which their mother has hollowed out beneath the snow. They are small (up to 1ft long), almost naked, and blind. By March they are sufficiently grown to venture outside the den and will start to eat carrion or fresh meat. However, they are often suckled at the same time for another year. The cubs quickly grow a dense undercoat of fur and a thick layer of blubber to protect them from the freezing temperatures

Arctic summer is at the disposal of the mother and cubs. If a mother should lose her cubs for any reason she may mate again immediately. Otherwise she will wait for another two or three years.

The young polar bear cubs appear to be full of play as they gambol in the summer sun, but in fact this is a time of education and of learning all the tricks of living which will see them through the ensuing years. They are taught how to find, stalk and kill seals, and how to swim. During these months the young bears are rapidly accumulating a thick layer of blubber and growing the dense underfur which helps to protect them even when swimming in the bitterly cold polar seas. Before their second birthdays they are on their own, and it is now that early training in survival becomes all important. If they can get through their second winter successfully they are well on the way to living to old age.

Male polar bears live solitary, nomadic lives, often migrating south over the floes in winter. During the brief mating season they sometimes fight viciously among themselves, but once this is over they leave the females and take no further part or interest in the upbringing or lives of their cubs. Occasionally inexperienced yearlings will band together, but these communities are not permanent.

Although it is primarily a hunter, the polar bear will feed on anything edible it happens upon; in this way, at least, it shows affinities to the terrestrial bears. All bears are scientifically described as plantigrade quadrupeds, which means that they walk on the soles of all four feet. Their characteristic shambling gait is caused by advancing both feet on the same side simultaneously. Terrestrial bears are so ponderous that they are unable to pursue fit animals and must usually rely for meat on chance encounters with wounded specimens. Otherwise they feed mostly on leaves, shoots, berries, nuts, roots and insects. Their bulk is enhanced by their long fur, but they are nevertheless enormously strong, and there are few animals that do not give them a wide berth.

Like most solitary, wandering animals, bears have not acquired many threat displays. Their reaction on being confronted by danger is either to flee or to attack. Today they are typically creatures of the colder northern forests, but in times past they had a wider range, and a species of the grizzly is still found as far south as Mexico. Bears define their personal territory (which may exceed 10 square miles) by the typical carnivore methods of scratching (marking) and leaving their scent on trees etc. on the boundary.

Bears are said to hibernate in the severe winter of the northern coniferous forests. This is not strictly true. They spend their winters in a deep sleep, denned-up in sheltered holes and caves, but unlike true hibernators their body temperature does not drop appreciably.

The giant panda
It was not until 1869 that the French naturalist Père David alerted the Western world to the existence of this familiar and strange-looking animal. Only in

1936 was a live specimen seen outside China—in the Chicago Zoological Park. Zoologists still argue about how the giant panda (*Ailuropoda melanoleuca*) should be classified among mammals. Originally it was thought to be a bear. Later both this species and the smaller red panda (*Ailurus fulgens*) were placed in a subdivision of the raccoon family. Very recently there have been moves to include the giant panda in the bear family again and leave the red panda in the raccoon family.

The Giant Panda can weigh over 300lbs and stand more than 5ft. Little is known about it, even in China, where it inhabits the cold, damp forests of the south-western and central regions. It lives exclusively on bamboo at an altitude of from 5000 to 10,000ft, and normally gathers the shoots at ground level, though it is said that it will climb trees, especially when in danger.

Generally regarded as one of the world's rarest mammals, it has been seen increasingly in Western zoos, where it has been presented as a gift from the Chinese government. Although it has bred in captivity, little is known about its breeding behaviour in the wild, though it is believed to produce only a single baby, probably every other year.

Another baffling question is the reason for its bold black and white markings. Many theories have been advanced, but none are satisfactory. However, the giant panda is totally protected in its native country and its endearing appearance has made it a favourite sight at those zoos lucky enough to possess it.

Tigers and lions

Of all true carnivores, only a few of the big cats have overcome the disadvantages of size and claimed its benefits. The tiger (*Panthera tigris*) and lion (*Panthera leo*) are prime examples. Both are of formidable size and yet both, in their very different ways, are astonishingly effective. These two species provide an excellent example of how quite closely related animals often employ radically dissimilar methods of solving the same problem—survival. The tiger is slightly the larger—measuring up to 13ft to the tip of its tail and weighing about 650lbs. Its range is scattered from the USSR and China in the north, through India and as far south as Java, Sumatra and Bali. The lion is now only found in Africa and in the Gir Forest of north-west India. (In the Ice Age its distribution extended to Britain in the west and Ceylon in the south. It was extinct in Europe by AD100.)

The tiger evolved as an inhabitant of the cold northern forests—where isolated populations still survive—and only in comparatively recent times spread south to the equator. Even the races from the warmer climates are happier in cold weather, which may help to dispel the surprise shown by visitors to zoos when they notice the obvious discomfort of a tiger in only moderate sunshine.

Like most cats tigers are rather solitary, guarding their huge territories (which may extend to 250 square miles) jealously, and preferring to hunt

Wolves are placental mammals. As with mono-tremes and marsupials the young are fed on the mother's milk, but the development of the baby takes place for a far longer period in the uterus, where it is attached to the wall by a placenta, a spongy mass of tissue through which the developing embryo obtains nourishment

alone. Unlike the lion, which will hunt by day or night, they are mainly nocturnal. They rely on stealth to capture their prey, either by stalking their victim or by lying low and waiting for the prey to approach. Their striped skins provide excellent camouflage, they have fine senses of sight and hearing (scent is less important), and are capable of explosive power. Once the prey (usually a large, hoofed animal) is within range, the tiger kills from behind, after perhaps no more than two or three gigantic bounds. Death is most often brought about by a strangulation bite to the neck, though smaller animals may have their necks broken. By such an attack and execution, tigers avoid the dangers presented by thrashing hooves and wounding horns. A single meal may consist of 40lbs of meat, and it has been estimated that an adult will make about thirty kills a year, representing some 6000lbs of animal. With a large kill, such as an ox or a buffalo, what is not eaten at the first sitting is often hidden and saved for future meals.

By these hunting methods the tiger has overcome its inability to pursue prey over long distances. The lion, which also has no stamina for the chase, has solved the problem in a completely different way. Almost alone among cats the lion is a social animal, living in 'prides' of variable sizes. By hunting co-operatively, in wolf fashion, if you like, the lion is able to pick off its panicked victims.

Despite its popular image as 'King of the Jungle', the lion is not a beast of the forest—although the remnant population in the Gir Sanctuary enjoys the protection of stands of poor quality teak trees. Everything about it, from its sandy colouring to its sluggish behaviour, suggests an animal adapted to warm, arid country with a minimum of shade. The males are impressive to see, with their shaggy manes and large heads, but they are lazy potentates, allowing the lionesses to do most of the hunting and the rearing of the cubs, while they content themselves with maintaining their quota of females and being first in when the prey is delivered.

A pride of lions is not quite the same as a harem, as there may well be more than one adult male in it. A pride is a nervous, tense society, comprised as it is of irritable and fierce killers, but the advantages of this gregariousness clearly outweigh its disadvantages. Cubs can be reared communally and food can be shared at a meal, with little of the potential wastage exhibited by tigers. A single hunting territory may cover 50 square miles, but this size is partly governed by the number of prey animals available. It has been calculated that in lion country there are likely to be 1000 herbivores for every three or four lions. Approximately half the prey taken will be wildebeeste, zebra species a sixth of the total, followed in decreasing order by Thomson's gazelle, buffalo, giraffe and impala. The quantities taken obviously vary, but in one study of a pride consisting of one adult male, two females and three cubs living in open country in East Africa, 219 herbivores were killed over a year.

Elephants

Elephants are the largest of all terrestrial animals. Once their relatives formed an extensive and numerous order, but now are reduced to two species: the African elephant (*Loxodonta africana*) and the Asiatic elephant (*Elaphas maximus*). The African elephant is slightly the larger of the two, a bull weighing more than 6 tons and standing over 11ft at the shoulder.

Perhaps the most remarkable of many features is the elephant's trunk, which is in reality the nose and upper lip combined. It has a variety of uses and is at the same time both tough and sensitive. While it can be used as a deadly weapon and as a sort of olfactory 'watchtower', its most consistent function is as a provider of food and drink. The trunk is used to grasp vegetation from the ground, from standing trees and shrubs, or from those which it has bulldozed through its colossal strength and weight. An elephant drinks about 40 gallons of water a day, sucking up 2 gallons at a time into its trunk and shooting it with incredible force into the open mouth—or it may squirt it over its back to cool its skin.

When one considers that one elephant may eat as much as 700lbs of vegetation a day and destroy as much again, the depradations caused by a herd are terrific. This is one of the reasons why those in reserves have to be kept in check either by culling or removing to less populated areas. The enormous damage they cause affects all other animals in the area, for elephants do not limit themselves to foliage, but will uproot trees to strip off the bark or chew the roots, and those trees not uprooted are often ringed and left to wither, making them perfect fuel for a major grass fire. In this way forest areas soon become treeless savannahs.

Before Africa became departmentalized, elephants seldom left forested country. They moved on and left the ravaged land in their wake to recover. Today, with most African elephants confined to national parks, they are forced to circle the same wasted areas, eventually turning them into useless dustbowls.

Elephants' digestion at first might seem inefficient, since much of their food passes through little altered. In fact however, they eat such vast quantities that by passing it through quickly they can eliminate the large quantities of indigestible fibrous matter and still assimilate enough of the nourishing parts—a wasteful method, perhaps, but effective.

Bull elephants' tusks, which can grow to 10ft, are in reality continuously growing upper incisors. Apart from using them as weapons against other bulls or common enemies, they can dig out water with them in times of drought or expose salt licks in the sodium-rich soil. In these two ways, at least, elephants improve their environment. Adults have no canine or premolar teeth, and the individual huge cheek teeth, which suffer much wear, are replaced six times in an elephant's life. When the last set is used up, the elephant may well starve to death, though many save themselves by taking up residence near rivers, where the vegetation is lush and tender.

Like other large mammals living in hot areas, the elephant needs to draw on every possible resource to lower its body heat. This is why so many enjoy wallowing in mud and water. (It is probably also the reason why hippopotamuses have adopted a river-bound existence.) Tusks are useful for excavating these mud wallows, and the massive external earflaps (each measuring up to 30 square feet) are liberally supplied with blood vessels, making effective cooling organs when flapped to and fro—the ears alone are capable of reducing the blood temperature by as much as 9°F.

When angry, an elephant will fan out its ears to emphasize its already formidable appearance, but its awesome size and tough hide has ensured that it is little preyed upon, and this is reflected in its normally placid temperament. When it is provoked, however, it is extremely dangerous—between 4 and $6\frac{1}{2}$ tons travelling at 30mph is a force to be avoided. Its normal progress is a sedate walk, making amazingly little sound as it pads around on its huge pillar-like legs, whose bones are aligned vertically to minimize the strain on the joints. The feet themselves are flexible pads which spread out under weight and act as a buffer between the animal and the hard ground.

Whales

Whales and their relations are the only mammals, with the possible exception of the sea otter, to have totally undertaken a marine existence, never on purpose venturing on dry land. They are superb examples of evolution and adaptation to a new environment. The most obvious hindrances to the successful invasion of the sea by warm-blooded, air-breathing, long-limbed mammals were heat retention, respiration and movement. When these problems were solved, the secret benefits of the sea began to reveal themselves. Whales do not merely tolerate the sea, they revel in it and the large whales have grown to a size never before equalled on this planet. Not even in the age of the giant dinosaurs was there anything remotely approaching the length and bulk of the baleen (whalebone) whales.

Only with difficulty can whales be recognized as mammals at all. The hindlimbs have disappeared and been replaced by powerful horizontal tail flukes, which are moved vertically. The forelimbs have become broad flippers. The nostrils have 'migrated' to the top of the head. Indeed, the whole shape of the head has been drastically modified and the entire body streamlined into a torpedo shape, encased in nourishing, warm blubber, in some cases as much as 16in. thick.

Whales (or cetaceans as they are called) are divided into two distinct groups—the toothed whales and the baleen whales. Most of the toothed whales are smallish, fish-catching dolphins—an exception is the sperm whale (*Physeter catodon*) which lives largely off giant squids in the depths of the oceans and grows to an outstanding size. In many ways this is most people's idea of a whale. It can reach a length of 65ft and a weight of about 50 tons. Its head, which takes up a third of its total length, is huge and barrel-shaped and contains vast quantities of a valuable fine oil known as spermaceti. This was the prize that early commercial whalers sought and which resulted in a mass persecution of sperm whales. But it is the baleen whales that have been the most severely hunted, and practically all the twelve species are now seriously threatened with extinction.

Even the sperm whale is something of a dwarf compared to the baleen whales. These have been abominably exploited, and the once universal blue whale (*Balaenoptera musculus*) is now extremely rare and rivets the attention more than any other animal on earth. Its total length can exceed 100ft and it may weigh as much as 135 tons (equal to 30 adult elephants!). There are only a few hundred of these magnificent animals left, and this may be too few to provide biological stability—in other words it may be a doomed species living on borrowed time.

For all their size and strength, whales are gentle and sociable. They display great sensitivity towards one another and have a fine sense of community welfare, helping newborn calves to the surface for their first all-important breath and protecting and nursing elderly comrades. Most whales travel in large groups (schools) and communicate by a wide variety of sounds, which in the ideal medium of water have a range of many miles.

It is one of the paradoxes of the animal kingdom that the largest of all animals should thrive on food as small as plankton. How does the blue whale manage to gather the two tons it needs every day? The baleen whale's mouth is a cavernous affair and the tongue alone can weigh as much as 4 tons—the weight of an average elephant. From either side of the upper jaw, in place of teeth, is suspended a series of fibrous, triangular plates (baleen) formed of keratin, a substance similar to our own nails. With its mouth agape, the whale swims slowly through the clouds of plankton, the unwanted water being expelled or strained out through the baleen by special muscles in the huge tongue or throat, depending on the species. The inner edge of the baleen is frayed to act as a sieve, filtering off the planktonic organisms, which are then swallowed.

Perhaps the most remarkable feature of whales is their ability to withstand prolonged dives to great depths. To find the giant squid, the sperm whale regularly has to descend some 3000ft and stay below for well over an hour. Such a prodigious feat is made possible by delicate modifications to its respiratory and metabolic apparatus. Among many refinements (all of which interact to maximum effect) that of the heart is the most important. When a whale dives ('sounds') the heartbeat automatically slows down. This helps to conserve oxygen, which is directed only to those organs such as the brain and central nervous system that require constant nourishment.

Apart from man, the only enemy of the great whales is a member of their own order—the fearsome killer whale (*Orcinus orca*). Killers are large dolphins which hunt in co-operative schools. Each can grow to 30ft and they are armed with strong conical teeth which interlock, providing an unbreakable grip on any animal they track down. They are helped, as are other whales, to 'see' in the gloomy depths by a system of echo-location—similar to that used by bats—by which they emit streams of ultrasonic sounds and decode each separate echo.

Rhinoceroses

The five existing species of rhinoceros are on the decline. We are so concerned today about conservation, that it is often forgotten that biological extinction, however sad, is a natural occurrence. However, conservation sometimes helps to slow down the natural process.

Rhinoceroses make up nearly one third of the remaining members of the ancient animal order of odd-toed ungulates, or Perissodactyla. Ungulates are the hoofed animals, and the odd-toed ungulates also include two other families: horses (including zebras) and tapirs. These three basic types of ungulates have survived for differing reasons. Horses are fast, aware and endowed with reasonable intelligence. Tapirs are secretive and retiring. Rhinoceroses are very large and well-armoured.

Though rhinos are said to be bad-tempered and dangerous (especially the black rhinoceros—*Diceros bicornis*), their nature is normally placid and gentle, and it may be their weak eyesight which makes them nervous of any sudden nearby movement. Their sense of hearing and smell are remarkably acute.

The largest of all rhinos is the African white rhinoceros (*Diceros simus*), second in weight only to the elephants among land mammals. An exceptionally large adult male may stand 6ft 6in. at the shoulder and weigh as much as 3 tons. This, the white or square-lipped rhinoceros, can easily be distinguished from the other African species, the black rhinoceros, by its lips. In the larger white rhino the lip is square, as befits a grazer, while in the black rhinoceros the upper lip is modified for browsing from trees and bushes—pointed with a prehensile tip. Both species have two horns, and in the white rhino these can grow enormously long. The record is $62\frac{1}{4}$in. for the front horn, and that for the black rhino is $53\frac{1}{2}$in. The great Indian rhinoceros (*Rhinoceros unicornis*), on the other hand, has a single horn, and its skin is heavily folded, especially round the neck, forequarters and hindquarters, giving it the appearance of an armour-plated tank. In size it falls between the two African species, standing a maximum of about 6ft at the shoulder and weighing over 2 tons. This species also has a prehensile upper lip, and it feeds mainly on tall reeds and grasses growing in swampy places. It possesses a pair of exceedingly sharp incisors in its lower jaw which are dangerous weapons when the animal is aroused.

Both the African species are dark grey, and the name 'white' probably derives from the Dutch word *wijd* (wide), referring to the shape of its lips.

The two other species of rhino are the Javan and Sumatran rhinoceroses (*Rhinoceros sondaicus* and *Didermocerus sumatrensis*). The Javan rhinoceros, though smaller by about 6in., closely resembles the great Indian rhinoceros, but its horn is only half the size of its larger relatives, and is absent in the female—other species have horns in both sexes. The Sumatran species is smaller than any other rhino (about 4ft 6in. at the shoulder and weighing seldom more than 1 ton) and unlike the two other Asiatic rhinos has two horns. It should be said here that the horns have been the downfall of rhinoceroses. They are not made of horn at all, but very closely matted hair. In Asia particularly they are reputed to have powerful aphrodisiac properties and for this reason the harmless mammals have been slaughtered in huge numbers to provide horns, which, when powdered, will sell at almost any price.

One of nature's more interesting manifestations is commensalism, when a partnership is formed between two unrelated organisms in which both receive some benefit. Rhinoceroses form an outstanding example of this phenomenon with birds such as oxpeckers and bee-eaters. The birds gain free transport and a wide variety of parasites off the backs of their hosts or insects disturbed by the rhinos' feet, while the mammals, handicapped by poor sight, are equipped with a keen-eyed early-warning system.

The black rhinoceros is the most abundant of the five species, but though its numbers may be as much as 10,000 it is still drastically threatened by poaching and by the shrinking of its habitat.

Camels

Camels, like rhinos, are ungulates—hoofed animals—but unlike them they are *even*-toed ungulates—Artiodactyla. They belong to the most widespread and successful of all orders of large mammals—the one from which we have acquired most of our beasts of burden. There are two living species of camel— the Bactrian (two-humped) camel (*Camelus bactrianus*) and the dromedary or Arabian camel (*Camelus dromedarius*). There are also four related and smaller cameloids in South America. These are the llama, vicuna, alpaca and the guanaco. The camel family, indeed, originated in North America and reached Asia and Africa via the land bridge where the Bering Strait now flows, 10 million years ago.

Though the Bactrian camel and the dromedary are obviously closely related, they have to tolerate quite opposite extremes of climate. The former lives in the extremely low temperatures of central Asia, and to aid it it grows a long shaggy brown coat every winter which it sheds in the summer; while the dromedary lives in the hot deserts of south-west Asia and north Africa. Both animals have been domesticated for thousands of years but there is also a wild subspecies of the Bactrian camel (*Camelus bactrianus ferus*), found in south-western Mongolia and north-western China. As might be expected it is smaller than the domesticated variety and less shaggy, but nevertheless is extremely hardy, migrating in the summer heat to the mountains to a height of 11,000ft and descending again to the desert during the winter.

The camel's humps are fat storage organs, providing food (not water). Water conservation is effected by a remarkable adaptation. In humans, body

The Bactrian camel (above) and the dromedary (below) are most commonly thought of as beasts of burden. (Some breeds of dromedary can carry loads of up to 1000lbs.) But they have other uses as domestic animals. The flesh is highly esteemed, the milk is rich and nutritious and the hair is used in many woven materials and also for paint brushes. Camel dung is a valuable fuel in the treeless deserts

heat is dissipated by the evaporation of perspiration. Were this to happen with camels, they would obviously have to consume large quantities of compensatory fluid. In fact a camel lacks the almost complete layer of subcutaneous fat found in humans and instead concentrates its fat in its hump. Human fat slows the loss of body heat, but a camel, which also lives at a higher body temperature than we do, absorbs heat more slowly, for its covering of well-ventilated hair acts as a barrier. Unlike ourselves a camel also tolerates great rises and falls of body temperature and will not even begin to sweat until it reaches a temperature of 40°C (104°F), by which time the day's heat is declining. At night a camel's temperature will drop to about 34°C (88°F), which means that the following day will be well advanced before the camel noticeably heats up. Such refined adaptations do not mean that a camel need not drink, simply that it makes the best use of whatever water is available.

Camels are unusual in being able to assimilate water rapidly. One specimen was found to drink over 20 gallons in less than ten minutes. Any other mammal of equivalent size would die through water intoxication after a prolonged period of extreme thirst, for the red corpuscles in the blood would absorb the water and rupture. A camel's red corpuscles, however, are egg-shaped and can quickly take up the water to become spherical.

While we can lose only about 15 per cent of our body weight through dehydration before the condition becomes fatal, a camel can lose as much as 30 per cent. In man, perspiring draws water from the blood, which then becomes thick and sluggish until the heart can no longer pump it fast enough to take the metabolic heat to the skin. A sudden rise in temperature is the result, followed by explosive heat death. But a camel loses most of its water from its tissues, not from its blood, and correspondingly when it replenishes its liquid reserves it is able to regain its former body-weight very quickly. One camel which lost 10 gallons of liquid only lost a pint and a half of this from its blood.

During the winter, camels are independent of drinking water for several months, as their bodies do not draw on water reserves for heat regulation purposes, and even in summer, when the vegetation is dry and the temperature higher, they may be able to last for two weeks, though the general pattern with domesticated camels is to ensure a replenishment of water every four days. Bactrian camels are able also to drink water too brackish for other animals. Obviously the distance a camel travels will have an effect on its need for water and 25 miles a day in the desert is a reasonable mean, though the swifter breeds may cover 60 miles a day without refreshment.

Externally the camel has several useful adaptations for its harsh life. The eyes of the dromedary are protected from sand and sun by a double row of long interlocking eyelashes, and the nostrils can be closed completely by muscles similar to those used by marine mammals. In a sandstorm a dromedary will fall to its knees and extend its neck along the sand with its nostrils closed until the storm is over. To take its weight when resting, there are pads of hard skin on the chest and legs and the broadness of the feet prevent it sinking into the sand. The Bactrian camel's feet are harder and calloused, enabling it to clamber about over the rocks and snow of its rugged home. Though the lips of the camel are very tender, they appear to be immune to the thorns of the dry scrub on which they often feed.

Camels are not very intelligent, and it is this factor, rather than any love they may have for their master, which has made them possible to domesticate and train.

The giraffe

The giraffe (*Giraffa camelopardalis*), at 18ft, is the tallest of all animals and certainly the most distinctive large African mammal. Curiously, the neck, which at first sight appears so strikingly long, is not abnormally so when compared to the length of the foreleg—the normal means of comparison. As in almost all mammals there are seven vertebrae. The relative length of the neck can be easily seen when the animal tries to drink, for to reach ground level it has either to crook its knees or splay the front legs wide apart. The backbone is very short and it is this that gives the animal its disproportionate appearance, and strangely high dorsal spines at the shoulder create an impression of forelegs much longer than hindlegs—which again is not so. The height of giraffes enable them to browse well above the level of other terrestrial animals and so avoid competition. One of their favourite foods is the leaves of acacia trees. Like camels, their sensitive lips seem to be immune from the long thorns.

For long the Giraffe was thought to be the sole representative of a unique family, but in 1901 the okapi (*Okapia johnstoni*) was discovered by Sir Harry Johnston in the Semliki Forest, Belgian Congo, and it is in these gloomy rain forests of central Africa that this primitive and secretive animal survives. Outwardly the giraffe and the okapi are not very alike. The okapi is a reddish brown with white horizontal stripes on rump and legs. It stands about 6ft 6in at the shoulder and lacks the two small external horns of the Giraffe. In some ways it is more horse-like than giraffe-like, and the name 'okapi' is a Pygmy one for 'donkey' or 'ass'. Anatomically there is no doubt, however, that it belongs to the giraffe family. Not much is known of the okapi's habits except that it lives in pairs in the thick forests. It was formerly thought to be nocturnal, but this misconception arose because of its skulking habits and sharp senses.

Giraffes are elegant animals and graceful movers. Several forms are recognized, such as the Masai giraffe and the reticulated giraffe, and these forms are differentiated by the patterns of their skins. Only the lion seriously preys on the giraffe, though even this powerful cat prefers to tackle antelopes, for the giraffe is a formidable adversary and can put its bony, horn-crowned skull to devastating effect or lash out with a tremendous kick of its hind legs.

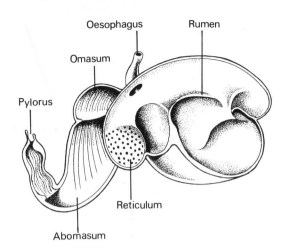

This diagrammatic drawing shows the four chambers of the ruminant stomach. Food passes down the oesophagus into the rumen, where cellulose in the vegetable food is broken down by bacteria. The 'cud' is brought back to the mouth and thoroughly chewed before again passing down the oesophagus and into the reticulum, from where it moves on to the omasum, which filters the food and extracts the water. In the final stage the proteins are broken down in the abomasum by the action of enzymes before further digestion takes place in the small intestine.

Other ruminants

Giraffes are ruminants, as are cattle, sheep, goats, deer, antelopes and camels. Most ruminants have horns, a reasonable defence, given their vulnerability to attack by predators. They rely largely for defence on speed, but this helps them little if they are cornered or ambushed at night. However, in many species the females are strangely hornless, and this is the case with all deer, except the reindeer (*Rangifer tarandus*). Deer shed their horns annually, but both sexes of the bovids (which include sheep, goats, oxen and antelopes) possess permanent horns. Ruminant horns are used both for display and threat purposes and for fighting, though it is true that several species rarely have recourse to real fights but use their appendages in mock battle to establish territory.

All ruminants are herbivorous, even-toed hoofed (ungulate) animals. Their most remarkable feature in common is their digestive system, from which they get their name. Out of 194 species of even-toed ungulates, 176 have three- or four-chambered stomachs. All deer, giraffes and bovids have four chambers and are called advanced ruminants. The chevrotains (*Tragulidae*)—also called mouse deer because of their small size and appearance (the lesser Malay chevrotain standing only 1ft at the shoulder)—have three-chambered stomachs and form an intermediate group between the advanced ruminants and the remainder of the order, which have the usual two-chambered stomach.

The digestive system of ruminants gives them two telling advantages over other herbivores: they can gather and swallow food quickly whenever it is available, without tiresome chewing (digestion taking place later when the animal is relaxing), and the maximum value can be obtained from the food, whatever its quality. As soon as the food is swallowed it passes into the first two chambers of the stomach (the rumen and reticulum), where bacterial action starts the digestive process. Later, at the animal's leisure, the food is brought up in small quantities and chewed throughly in the mouth—'chewing the cud'—until it is fine enough to be assimilated. It is then reswallowed and directed into the third and fourth chambers (the omasum and abomasum). There is no doubt that this sophisticated method of digestion has greatly helped ruminants to keep pace with evolution and reduce the pressures of predation.

Evading predation is the major shaping factor on ruminants' lives, and while the herding instinct of so many provides invaluable protection for the individual, others, like the pronghorn (*Antilocapra americana*—an antelope-like creature of North America but with branched horns which are shed annually—rely on sheer speed. It is the fastest mammal in North America if not in the world, for it can achieve speeds of 60mph for short distances and maintain half that speed for several miles. (The cheetah can reach a burst of speed of 70mph, but even over 500yds cannot maintain more than 44 mph.) The pronghorn is a curious animal, the only living species of the family Antilocapridae. Its horn appears to be part bovid, part cervid (deer), for though it is shed annually, there is an antelope-like bony core beneath the branched outer sheath.

Even-toed ungulates are sometimes referred to as cloven-hoofed. This means that the axis of the foot lies evenly between the third and fourth digit, while in the odd-toed ungulates the axis passes through the central digit. Even-toed ungulates are much more suited to mountainous country than odd-toed ones, for their cloven hoof gives them a far better grip on the slippery rocks.

The largest of all land mammals, the bull African elephant may tower 11ft at the shoulder, its 6½ tons supported on massive vertical tree-like legs. When charging, the 30 square-foot ears stand out from the head, the trunk is raised and the bull screams his rage with a fearful trumpeting. The curved ivory tusks can reach a length of 10ft. Unlike its Indian relative, the African elephant is not easily domesticated

Left Great herds of elephants can still be found in the wildlife reserves of East Africa. Confined largely to such gameparks, they circle and circle the areas, destroying the vegetation and creating vast dust bowls

Above Despite their protection in reserves, the numbers of elephants are threatened by the activities of ivory poachers who, risking severe punishments if caught, still carry on their lucrative trade

Right The male lion is a magnificent beast, but he is prepared to leave most of the hunting to the lionesses, though happy enough to take first pickings from any carcass

Above Well adapted to the Arctic wastes, polar bears are more carnivorous than their land-dwelling relatives. Seals are a favourite food and are often killed by a lethal cuff from the bear's paw as they emerge from their breathing holes in the ice. Surprisingly, polar bears are not particularly good swimmers, and have been known to be attacked by walruses and even mobbed by seals when in the water

Right A rare shot of a 40ft humpback whale as it leaves the water. The horizontal furrows on its throat enable the mouth cavity to expand when taking in scoops of its planktonic food. Fast swimmers, they are closely related to the blue whale, the largest animal ever to have lived

Above The great Indian rhino, one-horned and with a heavily folded skin, is becoming increasingly rare. Standing up to 6ft at the shoulder, it may weigh up to 2 tons

Right and far right The hippopotamus is the largest mammal to live predominantly in fresh water. When disturbed it will move from the shallows where it is basking and remain submerged for up to half an hour. Though rhinoceroses and hippopotamuses have some superficial resemblances, rhinos are related to horses and hippos to pigs

Smaller mammals

by Dr Michael Stoddart

The beaver

Few animals can have had such an effect on the landscape of the northern regions of the world than the beaver. Huge areas of rich meadowland, which were once forests, owe their existence to the beaver and its proverbial willingness to work. Evolution befits organisms to their environments in many ways and has gifted the beaver with the ability to modify its environment to suit itself. It is second only to man in this respect. Without this ability, life for the beaver would be impossible in its harsh habitats.

Beavers belong to the huge mammalian order of rodents. They are the largest species of the order found in the whole of the northern hemisphere— an adult male may weigh 40kg (85lbs). They have massive chisel-like front teeth which are used for gnawing through the trunks of quite large trees. They breed once a year, the litter of two to eight kittens being born in May or June. Life within the family is harmonious until the young are two years old, when they are driven out by their parents. In captivity they live for up to thirty years but in the wild, where they have to contend with wolves, pumas and wolverines, ten years is more likely.

Being essentially aquatic animals, beavers are adapted for life under water. Special valves close up their noses, ears and ano-genital regions and a specialized blood system allows them to dive for up to fifteen minutes at a time. These adaptations allow them to conduct their engineering works in deep water. Any size of stream is acceptable as a redevelopment site. First of all an island of twigs, mud and stones is built or enlarged. Over this a smooth dome of sticks, grass and mud is constructed. Care is taken to 'plaster' the outside with fine mud immediately before the first frosts of winter come. Once frozen, the structure is proof not only against the elements but against predators. The 'lodge' may be anything from 6ft in diameter to 40ft and is the heart of the colony. Here the beavers sleep, escape from enemies and, during spring, give birth. A chimney of narrow bore is left in the roof of the dome and is the only opening from the lodge to the outside. Entrance to and exit from the lodge is via two or three underwater passages. As the water level rises and falls with the spring snow melt and summer drought, these passages would be alternately flooded and exposed if it were not for the incredible dams built by these creatures. They are constructed downstream of the lodges, sometimes singly, sometimes in rows, and each may be anything from a few to several hundred yards long.

Ring-tailed lemurs, found only in Malagasy, are primitive members of the primates, the group to which man belongs. They forage for fruit and leaves at dawn and dusk. Unlike the smaller species of lemur they are active during the day. They are found only in the south-west of the island

The beaver, fabled for its incessant work, builds its 'lodge' from sticks and mud. Normally about 6ft across, though they may be larger, they are built on small islands or on the edges of ponds. They have a single room in them above water level, ventilated by a vertical air shaft, with entrances constructed underwater. The dams are created downstream of the lodge from sticks, logs, stones, mud, weeds and turf and can rise to 12ft and extend for 600yds. They keep the water level constant, so that the lodge is never in danger of flooding and the beavers always have enough depth of water to carry on their day to day activities unseen by enemies

The first stage in dam construction is to cut down many 40ft high trees. These are rolled and floated to where they are required, (and how the beavers know where this is nobody has any idea), and arranged parallel to the direction of flow. Boulders are dug up from the stream bed and piled on top. Not only are beavers compulsive workers but they are incredibly strong. Next come more felled trees and boulders and finally an interwoven wall of thin saplings and short logs. Sticks, leaves and mud are used for the final facing, both below and above the water level, and the end product is a very sound structure which is most effective in maintaining a constant water level. It used to be said that beavers showed great intelligence in dam building but this is not now thought to be true. The urge to build and repair is innate and never wanes. Thus, once finished, the residents set to with enlarging and repairing work, even if there is no need. Once the closest trees have been used, canals are dug to transport timber from up to 100 yards away. Feverish lumberjacking in autumn provides a store of maple, willow and cottonwood trunks at the dam. During winter when lush green vegetation is not available their barks sustain the colony. About one tree per day is needed by the colony of eight to ten animals.

As the construction sites grow too big, some dams are vacated and quickly fall into disrepair. As the water level falls the rich silt that has accumulated over many years becomes exposed. First water-loving plants, then grasses, herbs and shrubs appear, and consolidate the swamp. The result is a beaver meadow rich in plant and animal life. Soil analysis has shown that thousands of acres of the richest farmland in North America owe their existence to the beaver. A considerable part of the city of Montreal is built upon one such meadow.

Alas, man's greed for furs has decimated beaver populations. In North

America fair populations still exist in the remoter parts of Canada—elsewhere they are mostly restricted to game reserves. In Britain the last beavers disappeared during the twelfth century. They seem always to have been rare—even before the Norman Conquest in 1066, Howell Dda fixed the price of beaver skin at 120 pence while deer, wolf and fox skin fetched just eight pence! Beavers are still found in Scandinavia, on the Danube and the Rhone, in reserves in eastern Europe and throughout the USSR. Recent re-introductions into Switzerland and France are proving successful. Conservationists realize they can recreate and conserve one kind of natural environment with a high amenity value by re-introducing beavers. It is to be hoped that this realization will spread.

Mammalian reproduction

Since mammals are such successful creatures living in all kinds of habitats all over the world, it is safe to assume that they have an efficient system of reproduction. It is not strictly correct to say that all mammals give birth to live young, for the most primitive mammals, the monotremes (duckbilled platypus *Ornithorhynchus* and spiny anteaters *Tachyglossus* and *Zaglossus*) still lay eggs. Monotremes today are restricted to Australia and New Guinea and can hardly be described as successful, for all species are on the brink of extinction. Even the marsupials, the second of the three great groups of mammals are restricted in their range to Australasia with few exceptions. They do not lay eggs but their young are born at a very early stage in development and need to be protected and nourished in the pouch for many months. By far the majority of the 4200 odd species of mammals belong to the group called the placentals. In the placentals a system has developed whereby the young is kept inside the body of the female until its brain is quite highly developed. There can be no doubt that this has been of paramount importance in the success of the mammals throughout the world.

As far as its reproductive organs are concerned, there is little to distinguish the duckbilled platypus from the tortoise. Both have a single external opening to the genital system, urinary system and digestive system called a cloaca. Through the cloacal passage, a few days after mating, comes the platypus' two to three large yolky eggs. They are laid underground and are constantly guarded by the female. After hatching the young are transferred to a pouch on the belly and are fed off milk. Monotremes have no nipples so the babies must lick the milk off the sides of the pouch.

Marsupials have a slightly more advanced system and are thought by zoologists to mark a stepping-stone in the evolution of the placental system. A curious feature of the reproductive organs of female marsupials is that, after a short cloaca, there are two vaginas which lead upwards to two uterine horns. Correspondingly, males have a penis with two tips so that sperm can be deposited in each vagina. After about eight days of gestation the tiny marsupial is born. It has to climb, slither and crawl the few inches from the cloaca to the

Duckbilled platypuses are one of the most primitive mammals, from the Order of Monotremes, or egg-laying mammals. They have some other similarities to reptiles, but like mammals they can regulate their body temperature (though not so efficiently as higher mammals) and they suckle their young (though they have no teats and the milk seeps from the mammary glands). They have webbed feet and are found in streams in Tasmania and eastern Australia, where they feed on small aquatic animals

pouch (or marsupium, as it is properly called). Contrary to popular belief the mother apparently neither assists nor hinders the youngster in this journey. Zoologists think that the youngster is guided by odours it perceives from the pouch. Once safely there the sides of the lips grow together round the teat. To obtain milk it does not have to suck, for special muscles in the mother's abdomen gently pump out just the required quantity. Unlike the milk of placentals, which has a constant composition throughout the suckling period, the milk of marsupials gradually becomes richer as the youngster develops. Total emancipation from the pouch may take place long after weaning when the young animal may be up to ten months old.

Although there are very many species of placental mammals, so called because of the presence of a placenta (Latin for 'flat cake') which acts as an intermediary between mother and unborn baby, all reproduce in a very similar fashion. Most have distinct breeding seasons or seasons when the hormonal balance of both sexes allows and initiates courtship and mating behaviour, and allows fertilization and pregnancy to occur. Various factors influence the breeding season, such as rainfall, light conditions, food availability etc. The female reproductive cell, or egg, is made in the ovary and is fertilized by a sperm from the male somewhere on its journey down the fallopian tube. In some species, including man, the release of the egg (properly called 'ovulation') is quite spontaneous, but in others, including many rodents, the stimulation of coitus is necessary for its induction. Either way, once it is fertilized the egg finds itself in the uterus, or womb. Here it becomes embedded, or implanted, in the uterine wall and at the point of implantation a thick spongy wad of tissue develops. This is the placenta across which must pass oxygen and nutrients from the mother to the developing embryo. The important feature of a placenta is the close contact between foetal blood vessels and those of the mother in the uterine wall. The closer the relationship the better, since necessary transference can be effected all the faster.

In the normal course of mammalian reproduction, fertilization is followed by implantation and then development leading to birth. Many instances are now known where there is a delay in the process. Many bats mate in autumn but the sperm is stored in the vagina of the female all through hibernation. Actual fertilization occurs when the female bats wake up in the spring. A delay in implantation, but not in fertilization, occurs in many species of mammals, including seals, ferrets and, curiously since it is the only example in its huge order, the roe deer. The fertilized egg is held in the lumen of the uterus sometimes for several months before implantation and normal development proceed. A delay in the normal development is known to occur in some bats and rodents. Endocrinologists do not yet know how the hormonal control of these delays is effected. The result of a delay is that the young are born at the best time of year for their survival—when the food or climatic conditions are right. One of the most interesting questions currently facing reproductive physiologists is why the adults mate so far in advance of the time of birth that

The red kangaroo, at about 4ft 6in. high, is the largest of the family. It can reach a speed of 30mph, travelling in 12ft leaps, aided by its long and immensely powerful tail, which also acts as a support when the animal is stationary. When the baby kangaroo (only about $\frac{3}{4}$in. long) is born, it climbs up its mother's fur to the pouch and fixes itself to one of the four teats. After giving birth, the mother is soon re-mated, and within eight months a new baby is born. By this time the previous baby is able to live outside the pouch and so there is usually one young kangaroo on the ground and one in the pouch at the same time. In this drawing the baby inside the pouch, fixed to a teat, is about eight weeks old. Though still in a premature state, its distinctive long tail and hind legs are well advanced

a delay is necessary. There can be no doubt, however, that flexibilities of this kind have enabled mammals successfully to invade and colonize all parts of the world.

Desert mammals

Of all the regions in the world where mammals are found, deserts are quite definitely the least hospitable. The temperature extremes are not so unbearable as the shortage of water, although the viciously low winter temperatures of cold deserts act as a real deterrent to the presence of all but the hardiest mammals. The hot deserts are populated, in some places quite densely, with a variety of small mammals that are specially adapted to withstand the rigours of temperature and dehydration.

All mammals maintain a body temperature of about 37°C (99°F); if it rises much above then brain damage follows. Mammals from the temperate regions keep their body temperatures down in hot places by sweating and panting. Efficient as this system is, it necessitates the utilization of large quantities of body water and thus there is little place for it in an environment in which water is scarce. Amongst the large desert mammals, such as camels and donkeys, a physiological tolerance to rising temperature has been evolved. These animals do not start to sweat until their body temperatures reach about 40°C (104°F). Small mammals cannot tolerate high temperature but they can, and do, avoid it.

The desert by day is a bleak place with few signs of animal presence. By night one is astounded by the variety of animal life, from dung beetles foraging for nesting material, to desert foxes stalking lizards and desert rats. Most desert mammals live in burrows deep under the blazing surface of the sand. The tunnels of desert rats (*Dipus* spp., *Jaculus* spp.) may be six to twelve feet long, often with sharp hairpin bends in them. Straw plugs stop up the entrance and thus prevent moisture loss from the cool regions below. Deep in the nest the temperature is similar to that experienced during a summer day in northern Europe. By day desert mammals stay firmly at home and do not venture out until the cool of the evening, and so avoid the heat.

Desert predators, such as the fennec fox (*Fennecus zerda*), pace their day to match that of their prey. They have huge paws and use them to dig elaborate burrows. It is said that if a fennec is surprised in the open it burrows so rapidly that it almost seems to sink vertically into the ground! Deep within the burrow system their nests are cool and comfortable.

Some small desert mammals, like the antelope jack rabbit (*Lepus alleni*), from the desert regions of north western North America, do not live in burrows and must resort to finding whatever shade is available. During the day they remain quite inactive, moving only as much as is necessitated by the passage of the sun.

Small desert mammals may be able to avoid the extreme temperatures that occur but they cannot avoid the shortage of water. Carnivores, such as the fennec fox, have little problem, since their prey contains much water. The rodents, of which there are hundreds of desert species, feed on dry vegetation and seeds which contain very little water. Life is possible for them because they are adapted to derive all the water they require from their food. Most desert rodents never drink a single drop of water throughout their lives. Organic substances, such as fat, carbohydrate and protein, all contain hydrogen and oxygen atoms. When they are burned up by the body and metabolized during respiration, these atoms are freed and are able to combine to produce molecules of water. Every unit of carbohydrate yields three fifths of its weight of water; every unit of protein three-tenths of its weight of water and every unit of fat one and one-tenth of its weight of water. (Fats are very rich in hydrogen atoms which can combine with extra oxygen brought into the body during breathing.) Desert mammals produce very concentrated urine—up to five times as concentrated as in man—and so only need to use an indispensable minimum of water. Their kidneys are specially developed to remove excess water before the urine passes to the bladder.

Another major source of water loss in mammals is from the lungs. Inspired air may be almost dry while expired air is always saturated. Special condensation chambers in the nasal passages of desert mammals remove the water from expired air and return it to the body. The faeces of desert mammals are extremely dry, every last drop of water having been removed by the colon. Additionally desert rodents have no sweat glands and so are unable to lose water in perspiration.

It is not only water that is scarce for small desert mammals, for so also is their plant food. Deserts are characterized by xerophytic, or drought-resisting, plants, such as cacti and forbs (herbs other than grass). These plants make full use of the intermittent rains by blooming and setting seed within a very short space of time. Unless they exercised great prudence, desert rodents would experience occasional gluts of food followed by long famines. Most species have mastered the art of careful house-keeping and gather together huge stores of food when it is plentiful. If the collected seeds and pods were taken immediately into the cool humid conditions of the burrows they would rot. A system has evolved whereby the animals first of all bury the food in shallow scrapes where the heat of the sand will dry it out. After being turned and reburied several times the dried food is transported to special dry larders below ground. It can be stored there for many months.

Small mammals have adapted to life in the desert by avoiding the high temperatures and conserving what water is available to them. That they have been remarkably successful in doing this is evidenced by the high population densities often encountered.

Rodent population cycles

Every three to four years Norway lemming (*Lemmus lemmus*) populations over much of northern Europe grow very large and occupy areas where they

The fennec fox is a small animal of the Sahara, about 2ft long, including its 8in. tail. It is a nocturnal animal, as can be seen by its large eyes, and the bat-like ears are useful temperature regulators, allowing excess body heat to be dissipated to the surrounding air. The feet, too, are large and the pads covered with hair. They can be used to tunnel at great speeds. They live in cool nests deep below the surface

are not usually present. There may be so many animals that they migrate quite large distances to get away from their fellows. Their wanderings eventually take them to the coast and, as they launch themselves into the water in an attempt to swim to a new Utopia, almost all are drowned. The following year all appears to have returned to normal, for lemmings are seen nowhere but in the mountain tops. Such a repeatable and regular series of events is called a cycle. Population cycles of mammals occur throughout the northern regions of the world, but seldom in the tropics. In Europe and Asia cycles have a four-year periodicity and in North America a ten-year periodicity. This longer cycle is not based on a lemming species but on the snowshoe or varying hare (*Lepus canadensis*).

Rodent population cycles have been known since the time of the ancient Greeks. That grossly overworked God, Apollo, was given the extra chore of being responsible for ending mouse plagues. He was never successful for more than four years! In more recent times biologists have been engaged in identifying the ecological factors underlying cycles. An early suggestion was that North American species responded to the cyclical activity of sun spots, but is hard to explain why the animals have a ten-year and the sun spots an eleven-year periodicity. The true cause of cycles is still not known for certain, but biologists do not now think they are caused by just one factor. Current thinking interprets population cycles as the logical result of a complex series of environmental interactions.

Of fundamental importance is the observation that cycles occur only in the northern tundras. Life for mammals in these areas is harsh and only those specially adapted stand any chance of survival. Their chief adaptation is being able to breed extremely rapidly—Norway lemmings can mate before they are weaned and can produce ten litters of eight pups in each during their lifetime. When the winter snow is deep and long-lasting lemmings live and breed in the insulated space between it and the ground surface. If the spring is early, warm and dry, and new plant growth is fast, the populations soon become too big. Many lemmings are expelled to find a new home elsewhere. As successful breeding continues unabated, more and more refugees are cast out to take a chance on survival. As they wander they devour everything they encounter. Some manage to cross lakes and fjords, but most drown. By the start of winter a very few will have started new populations in uninhabited places, but the lion's share of the year's production will be dead.

As the lemming population grows, more and more predatory animals, such as wolverines (*Gulo gulo*), foxes (*Vulpes vulpes*) and eagle owls (*Bubo bubo*) move in for the easy pickings. Good food one year makes them produce more young next year and so a secondary cycle of predators lagging a year behind their prey is seen.

The combination of factors allowing rapid and sustained breeding, coupled with the time necessary for their plant food stocks to recover, seems not to re-occur until about four years later. Cycles are not seen in tropical regions,

The brown lemming is a rodent, related to voles, hamsters and musk rats. Its tail is short (about 1in.) and its body rather stumpy (5in.) and covered with thick fur. It is found in marshland throughout the Arctic circle, extending south of the circle in Scandinavia. The prey of stoats, Arctic foxes and snowy owls, it is a highly destructive creature, feeding on plants and tunnelling through the ground. Its population fluctuates at fairly regular intervals, as it literally eats itself out of house and home. Such an appetite means that, when the winter population is high, little food and practically no cover remains in the following summer. Many die of starvation and others are easy prey for predators. When food is scarce, the hunt for it leads the lemming into its famous mass migrations, when many perish by trying to swim over too wide a stretch of water in the search for food where constancy of climatic conditions does not allow different annual patterns of population to build up.

Sub-culture in Japanese monkeys

Human beings have long been interested in knowing how their various habits and cultures have spread as far and as quickly as they have. Recent studies on Japanese macaque monkeys (*Macaca fuscata*) have revealed not only that new cultures spring up quite often but, more importantly, how new habits spread.

Japanese macaques are highly social monkeys living in troops of up to sixty or more individuals. They are herbivores which feed on a wide range of plants and plant products such as roots, tubers, shoots and fruits. In winter they have to dig through the snow to reach edible plant matter. In 1953 workers from the Japanese Primates Research Centre noted that an eighteen-month-old female monkey started washing sweet potatoes before eating them. What motivated her to do this is a mystery, since unwashed sweet potatoes have been a favourite constituent of the diet for generations. The habit quickly spread to her special playmates and then to their mothers. Next the mothers passed the habit on to their infants. Last of all to join the craze were the rather unsocial males. Within three years the habit had spread to eleven troops, but even today not all troops exhibit this particular piece of sub-culture.

Laboratory studies show that the only food to which the monkeys react instinctively is milk. New foods, such as candy, are taken and eaten without hesitation by caged macaques, but quite a different response is seen in wild troops. When presented with candy, the first animals to try it are youngsters aged between two and three years. They are followed by their mothers, who pass the habit on to their infants. The more aloof males which show least interest in their young are the last to learn the habit. If by chance the adult males do learn a habit first, such as the recorded example of wheat eating by males, the habit spreads first to the adult females and then to the infants and juveniles. While it is not known how human habits and cultures spread, there is no evidence to suggest it was not along much the same lines.

Anteaters

During the long course of mammalian history several types of ant-eating mammal have been evolved. Ants and termites are found everywhere except polar regions so it is only natural that the ecological niche of anteater should be fully exploited. Although they rightly belong to different mammalian orders, the common way of life of ant-eating mammals dictates striking similarities. Thus from a functional viewpoint the anteaters can all be treated as if they belong to one group.

The basic equipment necessary for eating ants is a powerful pair of forepaws for tearing open rotten trees and termite mounds and a long sticky tongue for sweeping up the disturbed insects. Teeth are not necessary because the ants

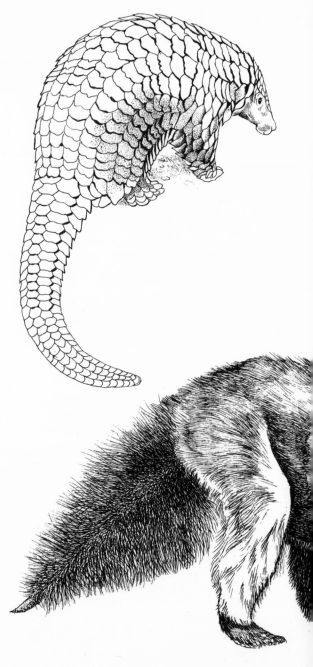

The seven species of pangolin or scaly anteater are found in the tropics from Africa to China. The Cape pangolin (pictured here) grows to about 3ft 6in. (including its heavy 18in. tail). All members of the family are covered with sharp overlapping scales, except on their undersides and muzzles, and can roll up into tight balls for defence. They feed on termites and ants which they lap up with their long sticky tongues, whose muscular roots are attached deep in their bodies

The spiny anteater or echidna of Australia is a monotreme, like the duckbilled platypus. Toothless, it uses its long sticky tongue to catch termites and ants, which are then crushed against the palate before being swallowed. When alarmed it can speedily dig itself into the ground using its broad, long-clawed feet

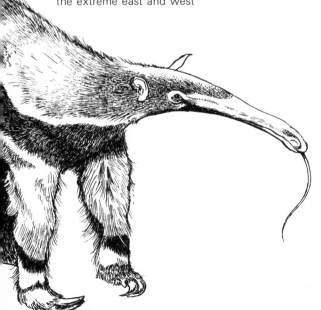

The South American giant anteater is related to the armadilloes and sloths. It has no teeth, but is able to sweep up termites with its long sticky tongue, after tearing open their nests with its powerful claws. It is a heavy animal, reaching a length of 7ft (including the long, bushy tail). Somewhat ungainly on the ground it can swim powerfully. This species is found throughout tropical South America, except in the extreme east and west

need only to be lightly crushed before being swallowed. Most anteaters seem to be impervious to the stings of enraged ants. Whether they truly do not feel the stings or whether they are biochemically immune to the poison is not known.

The spiny anteaters, or echidnas, of Australia and New Guinea (*Tachyglossus setosus*) are the most primitive mammals known to science. They are monotremes, or egg-laying mammals. They dig for insects with their shovel-like forepaws and have long, mucus-coated tongues for scooping up their prey. They are toothless.

The numbat, or marsupial anteater (*Myrmecobius fasciatus*) has about fifty teeth, but they are all degenerate and stump like. Little bigger than rats, numbats scrape around on the surface of the soil for ants and grubs. They are the only anteaters endemic to Australia and are not as highly specialized to this way of life as the New and Old World species.

South America is host to several types of anteater of which the giant anteater is the best known (*Myrmecophaga tridactyla*). It is six feet in length from the tip of its tube-like snout to the end of its bushy tail. Its long tongue is capable of sweeping up 500 ants at a time. The body fur is coloured black and white in a camouflage pattern and there is a marked 'V' pattern on the snout. This pattern was adopted by the British for camouflaging the barrels of their field guns in World War I.

The pangolin (*Manis temmincki*) is an African anteater which looks quite unlike a mammal because it is covered in scales. When it is curled up it is virtually free from all predatory attacks. Its tongue is over a foot long and controlled by long muscles which stretch back to its larynx. Like other anteaters it has immensely sharp and strong claws.

Marsupial 'mice'

During recent years ecologists have come to recognize that mammal life histories fall broadly into two categories. Some species, mostly small ones, have a short life span but breed frequently and have large litters. Others, mostly the large ones, live a long time but produce few offspring. One species with a curious life history strategy is the marsupial 'mouse' *Antechinus*. There are many types of these rather unspecialized pouch-bearing Australian mammals which live in deserts, arable lands and tropical rain forests. There is a pecisely timed breeding season each year which lasts about ten days. Those males that have survived the ten or eleven months since they were born all die within three weeks of mating. A few females also die but many survive until the following breeding season. This type of strategy—live many months, mate once, then die—is quite unique among small mammals. It is not clear exactly what causes death, but certain blood changes have been noticed in males at the time of mating. The blood becomes quite anaemic and it is likely the creatures die of acute anaemia. It has been suggested that the amount of testosterone, male sex hormone found in the blood in large quantities during the mating season, may play a part in initiating blood changes.

A few weeks after the mating season is over, the marsupial mouse population consists solely of pregnant females. The male youngsters they bear must, like their fathers before them, live to sire only one litter. Ecologists have no idea how such an arrangement evolved nor what are its particular advantages over the strategy normally adopted by small mammals.

Prairie dogs

Competing with farmers for land over huge areas of North America are prairie dogs, *Cynomys ludovicianus*. These are not dogs but large burrowing rodents, sometimes called prairie marmots. They once lived in colonies of millions of individuals but the introduction of new rodenticides is reducing their numbers. It used to be said that they shared their burrows happily with rattlesnakes because the latter were often found within the underground galleries of the former. The truth of the matter is that the snakes seek out the prairie dog homes for use as hibernation sites. Not unexpectedly the rodents decide that retreat is the best tactic to adopt and so the snakes effectively cause the rodents to enlarge their burrows.

Like so many rodents, prairie dogs are fanatical workers. This is a good attribute, because without constant engineering work the whole burrow systems are liable to flooding. When the rains come and the humidity inside the burrow rises the rodents scrape some soil together and mould it in the form of a small ridge round each of the hundreds of burrow entrances. They use their paws and blunt noses for this work. Then, when the floods come, the water merely washes round the little walls of compacted mud and the burrow system remains quite dry.

A colony of prairie dogs is called a 'town'. Each town is divided into several 'wards' which differ from one another by having dialectal variations in vocal and possibly also olfactory signals. Within each ward are several 'coteries'. Coteries are the basic social groups and consist of one mature male, his two or three wives and their still immature youngsters. During the spring there is much fighting between neighbouring males over who should lead which coterie. The victorious male in any encounter annexes for his own use the vanquished male's harem. Defeated males wander through the ward trying to win a new coterie, but if they fail they quickly become homeless and fall prey to black-footed ferrets or eagle owls.

As the young males in a coterie grow up they are driven out by their father and attempt to establish a new coterie on the outskirts of the town. Young females stay in the coterie and may even breed with their own fathers. Just before giving birth, pregnant females shut themselves away in dry, fur-lined breeding chambers. Their pups are born blind and naked, but grow quickly and are weaned in about five weeks. At the end of the summer, when the pups are almost full sized, it is not unusual for the mothers to leave their coteries and establish themselves at the periphery of the town whither the young males had gone some months previously. Just why they should do this is unclear. One suggestion is that this is an altruistic behaviour with the survival of the litter its object. When food is short one less mouth to feed could make all the difference between the litter starving or surviving. Alternatively it could be a mechanism for preventing too much inbreeding and subsequent weakening of the stock. Whatever the reason, there is no evidence to suggest that the continuing battle for survival between prairie dogs and farmers is nearing its conclusion.

Bats

If beavers are the civil engineering experts of the animal world, then the insect-eating bats are the electronic engineering experts. Insect-eating bats fly at night and not only navigate themselves about by radar but catch their prey with it as well. As if this specialization was not enough, there are one or two other curious bat characteristics. Perhaps the best known is the vampire bat—a tiny bat that feeds on the blood of sleeping mammals. Vampires are the only bats that can walk. They land close to their quarries and walk the last few yards. With their razor sharp eye-teeth they slice off a layer of skin, usually from the nose or the ear. As the blood wells up in the wound they lap it up. Over much of the tropical world vampires carry rabies in their saliva and thus are the subject of huge irradication programmes. Equally fascinating in its habits is the fish-eating bat from Central America. This is a large bat which scoops fish out of the waters of rivers. It is thought it uses a form of radar to find where the schools are, but how it manages to get the beam of sound through the water and back is not known. Perhaps it hears the school break the water. These bats are efficient fishermen and seldom fail to catch a four-inch fish in a few minutes.

What we know about the navigational abilities of the insect-eating bats is quite astounding. Their system of radar is far superior to the most advanced military guidance systems we have yet designed. The stream of sound pulses is produced in the larynx, or voice-box. Some species produce a noise audible to us but in most cases it is ultrasonic. The pulses have a frequency of between 15 KHz and 150 KHz (man can hear up to about 20 KHz) with wavelengths of between 2.2-0.22cm. Each burst of sound lasts for as long as 60 milliseconds or as little as 0.25 milliseconds. Since these bursts last for such a short time and follow one another very quickly, the radar is capable of extreme sensitivity and resolution. Bats have been shown to detect, and avoid, wires just 0.1mm ($\frac{1}{250}$in.) thick. Curiously enough, once a bat knows its way about its territory it shuts off its radar and flies blind. Bats kept in private houses can be completely fooled by a closed door which is normally kept open!

Some bats sing their notes while others hum them through their noses. These latter, the horse-shoe and spear-nosed bats, have grotesque flaps of skin dividing the nostrils and forming directional megaphones. The singing bats have no need for these structures since the sounds come from an open mouth. Both types have greatly enlarged ears—in one common European species the ear is as long as the body—often with enlarged ridges and tongue-like structures. Not only are these to pick up the echo of the note but they assist in the precise location of moving prey by themselves moving. The ear flaps of horse-shoe bats wave backwards and forwards as often as fifty times every second.

Human radar engineers know that two different types of sound burst are required for measuring distance and speed of movement. Distance measurement is best made by a note which drops in pitch through two to three octaves, called a 'sweep'. Speed of movement is best measured by listening to the change in pitch of a long constant note made by a body as it draws nearer to or further away from the listener. This pitch change, known as the Doppler effect, is self-evident when witnessing an express train going past. In order to compute the position and speed of a tiny flying insect, bats must utilize both kinds of sound production. As they home in for a kill, the rate of pulse production rises and the duration of the pulses shortens.

Mostly insects are scooped up in a wing, tossed into the tail membrane and then eaten. They are seldom caught directly in the mouth. Some insects, several types of moths and lacewings have developed a means of detecting bats so they can take evasive action and avoid being eaten. One group of moths has even developed a sound-emitting system of its own which serves to dissuade the bats from attack and thus functions much like the bright yellow and black warning stripes on bodies of distasteful insects.

Most insect-eating bats live in the tropical world where there are active insects flying the whole year round. All species spend the winter in hibernation in caves and hollow trees. In the tropics they may remain active the year round. One of the seven wonders of the biologists' world is to witness the activity of bats and swiftlets at the Gomantou caves in North Borneo. The caves are occupied at night by swiftlets (*Collocalia spp.*) and in the day by free-tailed bats (*Myctinomus plicatus*). The changeover period lasts several hours each dawn and dusk. Collisions are, apparently, extremely rare.

The honey badger

A co-operative relationship between two species of animal, in which both benefit from the association, is called commensalism. One of the most fascinating examples we know concerns the honey badger (*Mellivora capensis*) (or ratel as it is sometimes called) and the honey guide bird (*Indicator indicator*). The partnership is not essential for life, because honey badgers live quite successfully without the birds in India. It is only in tropical Africa that the relationship flourishes. As the name implies, honey badgers feed on honey as well as bees and grubs. Having an immensely thick, sting-proof coat and long, curved claws, the honey badger is well suited for tearing open

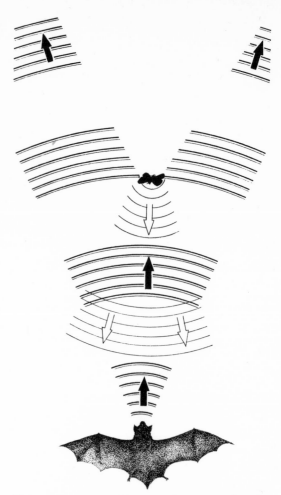

The sonar system of bats works on a series of very high-pitched squeaks, emitted at about 1/10 second intervals. They can range up to 100,000 cycles per second (humans can hear up to 20,000 cycles per second). When these sound pulses collide with an object they are reflected to the bat as echoes, so that the bat can both avoid obstacles and track down its food. Bats with 'nose-leaves' use them to direct their sound, while in other species it is the shape of the mouth which determines the diameter of the beam of sound emitted

rotten trees which house wild bee nests. The honey guide locates the bee tree and then leads the badger to it through the forest. After the latter has torn the wood aside both feed on the contents of the nest. Since the honey bird eats only the wax combs, it is easy to see how successful the partnership is. The remarkable thing about this relationship is that the badger tries to inform the bird of its wishes by putting itself plainly in sight of the bird and uttering a series of soft whistles. Getting the message, the bird flies off to search for a nest. Finding one it returns to where it left the badger and utters a rapid and characteristic series of excited sounds. These it keeps up while flying a few yards in front of the badger until it reaches the nest. The cries are not specifically designed for the badger's ears and often other animals, including man, are attracted by the commotion.

Honey badgers are only distantly related to the true badgers of northern temperate lands and are perhaps best allied with the wolverine. They are bad tempered and extremely vicious, attacking anything that crosses their path. The pleasant *quid pro quo* relationship badgers have with the honey guide is thus all the more curious.

Dogs and foxes

Although dogs and foxes are superficially much alike they are ecologically and behaviourally quite distinct. Together with the medium and large cats they fill the niche of major predators and sometimes specialize on prey larger than themselves. Foxes tend to live solitary lives and feed on small birds, rodents and insects. Dogs, on the other hand, are gregarious and sometimes live in packs of several dozen. Being aggressive animals by nature, the behaviour of dogs shows many advancements over the behaviour of foxes, particularly in that aggression can be sublimated in the interest of group harmony.

Foxes are smaller than dogs and the smallest, the desert fox or fennec (*Fennecus zerda*), weighs only three to four pounds when adult. An adult red fox (*Vulpes vulpes*) from Europe or North America weighs around 50 pounds. By comparison a male wolf may weigh 120 pounds. A huge range of coat colours is seen in both types, depending on whether the habitat is woodland, desert or arctic waste—from white through greys, reds, yellows, oranges to tortoiseshell and black. Since both dogs and foxes have very keen hearing their ears are large and erect to catch the slightest sound. Their long noses indicate a keen sense of smell.

Both foxes and dogs are widely distributed over the globe. They occupy all main habitat types except the air and the forest canopy. In this respect they echo the success of the rodents, and for much the same reason. Fox and dog populations contain a great deal of genetic diversity which allows them to adapt to changing conditions. It was this untapped diversity present in ancestral wolves that allowed our forebears to develop more than 100 different breeds of domestic dog. While the wolf is undoubtedly the main source of genes in domestic dogs zoologists think that jackal, coyote and dingo blood may have been introduced from time to time. It is hard to accept that the genes for a chihuahua are present in a Canadian timberwolf!

Foxes are just as adaptable, but their genes have never been introduced into domestic breeds. This is because their way of life, stealth and cunning, are not traits useful to man. But when myxomatosis decimated the rabbit population of England and Wales in the 1950s red foxes, their major predators, switched their feeding habits and started a suburban scavenging life. The huge amount of food scraps thrown into dustbins each day by humans offered a whole new existence to the fox. Foxes are intelligent creatures and even the heaviest dustbin lid was no match for their skill. After myxomatosis passed, the rabbit population slowly started to increase, but the fox did not revert to its earlier way of life in any great numbers. Today the spread of the urban fox is one of the greatest problems facing British pest control officers.

Some types of dog, notably the hunting dogs, live all their lives in a pack and stand little chance outside it. Others, like the timberwolf, can live singly or in packs, as the food conditions warrant. In northern Canada it is quite usual for wolves to live either alone or in family groups in the summer when a good living can be made off birds, snowshoe hares, beavers and deer fawns. During spring, the litters of two to six pups are born in a hole in the ground dug by the mother. Three months later the pups are weaned and stay with their parents during the winter. As winter advances family groups join together in packs of up to thirty. This is essentially a hunting co-operative but it has a rigid social structure. A single male and female reign over the others and take on themselves the important job of patrolling the territory boundary. The territory may be twenty square miles in area, and is hotly guarded against intruders, whether they be single wolves or neighbouring packs. Domestically, the two leaders settle disputes between pack members so that a fragile peace is maintained. The pack hunts daily for elk, moose and other deer and displays complicated manoeuvres for enmeshing its quarry. There is no doubt there is more food for everyone as members of a pack than there would be if the pack broke up and individuals hunted on their own. If the winter is excessively hard and the food supply runs out, the most lowly members of the pack's hierarchy are cast out. Usually outcasts soon die but they may hang around near the pack for a while in the hope of scavenging a meal from a stripped carcass. The hierarchy is established and maintained by a complicated series of tail semaphores, ear, mouth and body movements of the wolves, as well as by their vocalizations.

In sharply marked contrast, almost no social activity is seen during the winter between foxes. As spring advances and with it the breeding season, individuals seek out their mates. Foxes tend to be monogamous and to pair each year with their mates from the year before. Vixens deposit highly scented urine marks at obvious places in their home ranges and indulge in much howling to attract a mate. At this time of year dog foxes travel huge distances in search of a vixen. Once found, mating takes place and then the pair sets off to dig out a den in a safe place. Gestation lasts nine weeks and a litter may have up to ten cubs. During the early weeks there is much play between cubs which serves to establish a hierarchy or peck order. As one of the adults returns to the den with some food there is a complicated licking—greeting—behaviour. Between two and three months of age the cubs join their parents on hunting trips and by the start of the autumn the family splits up, each to go his or her own way in savage intolerance of all others until the next spring.

Rodents

If one considers the success of a group of animals in terms of number of species then there is no doubt that the gnawing animals, the rodents, are the most successful of all the mammals. Nobody knows for certain just how many species there are but it is somewhere between a third and a half of the total for mammals. While some species are quite distinctive, like the pig-sized South American capybara (*Hydrochoerus*) and the porcupine (*Hystrix*), by far the majority are tiny, rat and mouse-like creatures living in burrows beneath the soil's surface or in the litter just above it. There are probably many species yet to be discovered. Rodents are ubiquitous creatures, being found in every continent except Antarctica. They are the only land-dwelling placental mammals that ever reached Australia and were able to exist in competition with the earlier established marsupials. No other group of mammal occupies such a wide range of habitats. Rodents are found from the extreme northern edge of the Arctic tundra to well above the tree line in high mountains, from the centres of man's cities to the desert in which there is no visible vegetation, from the sands of East Africa so hot they will blister human skin to the subzero temperatures of our northern wastelands. They live in water, swamp, scrub, grassland, farmland and even in the canopy formed by gigantic tropical trees. Just after World War II rodents were destroying more grain than was allocated by the US for her entire foreign aid programme. Wherever man has planted crops he has been fighting rodents. Wherever he stores grain he continues the battle. It would seem likely that the successor to man as dominant animal on the globe will be a rodent.

Why are rodents so successful? This is not an easy question to answer for success depends upon the complex interaction of many factors. Perhaps of most importance is the ability of rodents to adapt to new and changing conditions. Genetically speaking they are young mammals which have not lost their initial genetic variability. New conditions can be adapted to so life can continue as before. A good example of this concerns the history of Warfarin—a rodenticide developed by the Wisconsin Alumni Research Institute and used since 1949. Warfarin acts by interrupting the body's mechanism for forming blood clots. Since tiny haemorrhages occur all the time in tiny blood vessels throughout the body, the clotting mechanism is in constant use. In its early trials and applications Warfarin was very successful but gradually its effectiveness waned. In 1960 the first truly resistant rats were recognized in Scotland. Today pockets of resistance are found in many parts of Britain and Western Europe. Intensive research has shown that certain genetic changes have occurred in resistant strains. From a rodent control point of view this is rather alarming since it would appear that a similar resistance could be developed to counter any newly researched poisons.

A second factor in the success of rodents, and linked to the first, is size and breeding rate. All rodents are small and are thus able to grow to adult size, and hence breeding size, very quickly. The black rat (*Rattus rattus*) reaches maturity at three to four months. Gestation in small rodents is about three weeks and a litter consists of up to eight pups. Young black rats are weaned and capable of independent life at about four weeks. Although average individuals live perhaps no more than two years, eight or nine litters can be produced in this time. In the tropics there is no period when breeding does not occur. This fantastic output means that there is a very high genetic turnover or, to put it another way, many different genetic combinations can be tried out. Should the environment change, as with the sudden appearance of Warfarin, the chances are high that a preadapted type already exists. This high production rate also means that unexpected food resources can be quickly utilized as soon as they occur, as every grain warehouse manager knows to his cost!

The third factor concerns the ability of rodents to eat almost anything. All species are herbivores but some prefer a more omnivorous diet. Green plants cannot be digested by any animal, since none produces the enzymes necessary for breaking down cellulose. Herbivores keep a colony of bacteria somewhere in their gut whose function it is to break down the long cellulose molecules. In rodents the colony lives in the caecum, of which our appendix is the equivalent. This is far down the gut and well below the stomach where digestion takes place. Therefore a system of passing the food through the gut twice, or refection, has been developed. Once it has been acted upon by the bacteria the food is passed as a series of large soft droppings. These are immediately reingested and pass straight to the stomach for digestion. Refection is not unique to rodents, however, for it also occurs in their closest relatives, the rabbits and hares.

The more that is discovered about rodents, the more we see them as a threat to our way of life. Not only do they compete with us for food, but they carry diseases like bubonic plague and rheumatic fever. Their intelligence and ability to learn make them perhaps the greatest force to be reckoned with that civilized man has ever, or is likely ever, to encounter. It seems certain that the world food problem would be far less severe without food stocks having first to satisfy the prodigious appetites of millions of rodents.

The Japanese macaques are the most northerly of all monkeys, and live above the 5000ft line. Here, with their heads matted with snow, they are bathing in the hot springs which abound in the area

Above The fastest sprinter in the world (up to 70mph), the cheetah hunts small antelopes

Left and below Leaping from an acacia tree, a leopard gives chase to a screaming baboon, meeting a final desperate gesture of defiance before the pounce and the kill

Above There are only seven species of pangolin, or scaly anteater, all found in Africa and Asia. This Cape pangolin shows the typical sharp-edged overlapping scales, used both for attack and defence, and the muscular, wormlike tongue, with which it collects ants and termites. When attacked, pangolins roll up into a ball with their scales projecting

Right One of the prettiest deer in North America, and one of the commonest, the white-tailed or Virginian deer is an animal of light woods and thickets. When disturbed it erects its tail, whose white underside serves as a warning signal. The dappled coat of this fawn is a perfect camouflage in the light and shade of the woods

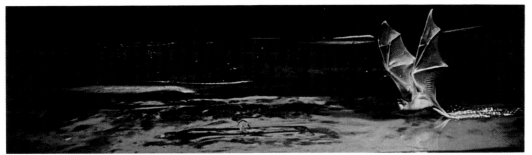

Left The huge ears of bats enable them to pick up echoes from their ultrasonic (up to 100,000 cycles per second) squeaks. By this echolocation they both detect their prey and avoid obstacles. Some species have appendages on their noses (nose leaves) which help to direct the squeaks

Above and below A remarkable photographic sequence of a fisheating bat, which catches its prey in its claws from the surface of sea and fresh water and eats it in flight. Fisheating bats are found from the Gulf of California south to Brazil

Birds

by Robert Burton

Introduction to birds

Birds are the most successful group of warm blooded animals alive today. After the Age of Reptiles, when dinosaurs ruled the world, there came the Age of Mammals which reached its zenith some 50 million years ago. Birds started to proliferate shortly after the mammals and have now overtaken them. There are 8000 species of birds currently in existence but only 4000 of mammals. So we may say that we are living in the Age of Birds. Every group of animals has at one time suddenly branched out, diversifying in form and habit. For birds the process is still proceeding at a steady pace, although it is almost too slow to measure by human terms.

When considering aspects of birds' lives in detail it is worth remembering that their success has depended on several attributes that have allowed them to take to the air. Each attribute was vital for the transformation into an efficient flying animal. Birds are feathered animals. Their feathers, which evolved from the scales of reptiles, are the key to their success. When birds first appeared, their feathers gave them an advantage over their reptilian forebears. By trapping an insulating layer of air next to the body, birds could become warm-blooded. This meant that they could be active at all times, whereas reptiles are sluggish at low temperatures. Feathers also play a part in the social life of birds. Brilliant colours and ornate plumes are used as signals in courtship, while a drab plumage helps to conceal them from predators, but feathers are, above all, necessary for flight. The coat of body feathers helps to give the bird its streamlined shape and the long wing and tail feathers provide the flight surfaces.

A suitable wing surface and warm-bloodedness, both provided by the feathers, are the two essentials for flight, but nearly every other part of the body is adapted to this end. An efficient digestion provides the energy required for flight and for keeping the body warm, and the high body temperature speeds the action of nerves and muscles. The huge breast muscles, anchored to the keel, or sternum, of the breastbone provide the power for flight, but the rest of the body has been made as light as possible. The bones are hollow and some have fused together to give strength with lightness. The teeth have been replaced by a horny beak or bill. The egg-laying habits of reptilian ancestors have been retained to avoid the necessity of carrying developing young within the body.

The limitations placed on birds by the necessities of flight have resulted in a generally uniform body plan. Yet, in outward appearance, birds are diverse.

The long-eared owl is found throughout the northern forests of the world. In this threatening display the long tufts (not ears at all), which give it its common name, can be clearly seen. As with all birds of prey, the eyes are large and face forward, giving binocular sight, and to compensate for this narrowed field of vision the neck is extremely flexible and, indeed, the head can almost be turned upside down

The Andean condor has the greatest wingspan (up to 12ft 6in.) of all flying birds. It is a vulture of the high mountains of Chile and Peru, where it lives between the 9000 and 16,000ft zone, feeding chiefly on carrion, though it will also kill the young of sheep and goats. Males (as shown here) have a fleshy crest on their heads. The general colouring is a striking contrast of black and white

They range in size from the tiny bee hummingbird of Cuba, 2½in. long including an outsize bill, to the 8ft ostrich. In shape, there is a range from the long-necked, long-legged herons to the almost legless, neckless, streamlined swifts. The variety of birds is a reflection of the rapid evolution of forms suited to different ways of life. Flight has allowed birds almost unlimited freedom of movement. They can spread through the layers of a tropical forest to exploit the wealth of fruits, seeds and insect life, and seabirds can breed in one place and travel swiftly to choice feeding grounds many miles away. Barriers that limit the movements of landbound animals have not hindered birds. They have spread to remote oceanic islands and migration has let many birds exploit temporary supplies of food. Without migration, temperate countries would never see the swallows, swifts and warblers that rely on small insects for food.

The mechanism of variation is the process of adaptive radiation, that is the evolution of a variety of forms with different habits from a common ancestor. The familiar example of adaptive radiation in birds is that of Darwin's finches of the Galapagos Islands. The thirteen species of this family originated from an ancestor that arrived from the mainland of South America, 600 miles to the east. In the isolation of the islands, without competition from other birds, Darwin's finches have developed a variety of feeding habits. There is a warbler-finch with a thin bill for catching insects; others crack seeds with conical bills or eat soft fruits, and the woodpecker-finch used cactus spines to tease insects out of crevices.

The feeding habits of Darwin's finches are intimately bound up with the size and shape of the bill. This can be seen in a wider context. Flesh-eating

Accurate data on animal speeds is notoriously difficult to collect, but it has been claimed that the spine-tailed swift has been recorded flying at 219mph. If this is so, it would make it by far the fastest bird. More conservatively, it can certainly maintain speeds of 60mph on its long wings and with its streamlined body. The spines on this swift's tail are modified tail feathers, which enable the swift to cling vertically to tree trunks. All swifts have tiny bills, are practically neckless and have legs so short that they cannot take off from the ground

birds, the hawks and owls, have hooked bills; finches have stout, conical bills for cracking seeds, and the flamingo's bristle-fringed bill strains minute algae from lake water. The diversity of birds is, therefore, based on the exploitation of different foods. While the bill plays an important part in feeding, the wings, feet and senses all have a role and add to the variety of birds. Thus, we have the slender wings of swifts that chase flying insects, the broad wings of the soaring eagle, the webbed feet of a duck or penguin and the acute hearing of an owl hunting in the dark. However, the importance of behaviour in bird life must not be overlooked. On the whole, a bird's life is organized by instinct. Swallows migrate to Africa and robins build neat nests by instinct. Food is found and predators are avoided by patterns of behaviour passed from generation to generation that have evolved to serve the bird to the best advantage. Learning is used principally as a fine adjustment whereby the bird adapts its instinctive behaviour to fit its particular circumstances. In the following sections, we shall see how birds have found ways of finding food and rearing families that allow them to exploit certain habitats or food supplies without coming into conflict or competition with their fellows. A massive change in anatomy turned the birds into efficient flying machines but an almost infinite variety of behavioural adaptions and modifications allowed them to exploit their flying powers to the greatest advantage and so become such a successful group of animals.

The lengthy courtship of albatrosses

There is a natural antagonism between animals that keeps them apart, as demonstrated by swallows neatly spaced along a telegraph wire. The prime aim in courtship is to overcome this antagonism and bring the sexes together; the male has to attract the female and she has to approach submissively so as to avert his aggressive instincts. For many birds, coming together may take only a few days, but for the long-lived albatrosses courtship extends over an adolescent period lasting several years.

Young wandering albatrosses return to their breeding grounds on the cold, windswept islands of the subantarctic when three or four years old. When they first leave the nest, their plumage is chocolate brown but it gradually becomes paler until, when fully adult at 18-20 years, it is pure white except on the wing-tips. Males lose the brown feathers more quickly which makes it quite easy to distinguish the sexes.

The courtship of the wandering albatross at first resembles a barn dance. A group of about six birds of both sexes gathers and they display to each other in turn, wheeling about with their wings spread. Individuals come and go frequently, but the first social contacts have been made.

The next stage is for the males to establish small territories where they display at the females who circle overhead. The display is spectacular. The albatross spreads its 10-11ft wingspan, points its bill at the sky and emits a shrill, screaming cry. If interested, a female will land and approach the male.

Some courtship takes place, then she leaves to investigate another male. Over a period, the female spends more time with each male until she starts to visit one regularly. They then undergo an 'engagement' and eventually the union is consummated and a single white egg is laid. By this time the birds are nine years or older. The pair will now stay faithful until one dies.

It is easy and amusing to draw a parallel with the sequence of human courtship, where there is also a period of trying out prospective partners before making a final decision. What we do not know is how albatrosses, and perhaps humans, decide who will make a suitable mate.

Feeding helps togetherness

After male and female become used to one another and have formed a partnership, the bond between them must be maintained while they raise their family. Among several groups of birds, courtship-feeding is part of the machinery for pair-bond maintenance. The male brings food to his mate which she accepts in a manner very like that of a chick begging for food. A hen greenfinch or linnet approaches the cock and crouches low with fluttering wings and uttering begging calls. The cock draws himself up to full height and lowers his bill to drop food into her mouth. In the gull family, too, the female reverts to infantile behaviour by approaching the male with head held low and neck withdrawn.

Courtship-feeding has more than the psychological function of maintaining the pair-bond. While forming eggs within her body, a female bird is under the physical strain of having to supply ingredients for the manufacture of the yolk, albumen and shell. It is clearly an advantage if the male can save her the effort of having to find all her own food. After the eggs have been laid it is the custom in several species for the female to undertake most, if not all, of the incubating duties. There is a division of labour in which she devotes her whole attention to the care of the clutch while the male forages. Male birds of prey rarely, if ever, undertake incubation duties and the female is dependent on being fed. If, however, there is a food shortage and the male cannot supply both their needs, she is forced to hunt for herself. There is then a chance of the nest being robbed in her absence, so that in years of food shortage birds of prey are often less successful at breeding.

The psychological value of courtship-feeding has been demonstrated experimentally in the common tern, which breeds on both sides of the Atlantic. When feeding their mates or chicks, terns carry small fishes or shrimps in their bills. It is quite an easy matter to identify the kinds of food being brought back to the colony and even to estimate the size of each item by comparing it with the size of the tern's bill. If the terns can be identified by means of colour rings, the breeding success of an individual male can be compared with the amount of food it brings during the few critical days during which its mate is forming eggs.

In a study carried out at Bird Island, Massachusetts, it was found that the amount of food fed to the female, who rarely feeds herself during the egg-laying period, affects the weight of the clutch. Terns lay two or three eggs and the third to be laid is often smaller than the others. The ability of the male tern to bring food for his mate significantly affects the size of the third egg and consequently its chances of hatching. There is also a correlation between courtship-feeding and successful fledging of the third chick, because the male is responsible for most of the feeding of the chicks during the first few days of their lives.

The courtship-feeding of terns is an illustration of natural selection at work. The males that are best at fishing raise the most chicks. So not only do the fittest terns survive, they produce the most offspring and these will probably inherit their fathers' qualities.

The Bird Island study also showed that female terns may be able to recognize which are the best males. The first stage of courtship consists of male terns carrying fish from female to female and displaying to each. Next, the pair is formed and courtship feeding begins, and in the third stage, the female remains in the territory and receives all her food from her mate. At each stage she has the opportunity to weigh up her mate's potential as a provider for herself and the family and there is evidence that she may leave him if not satisfied.

Competition for females

Sage grouse live in the high prairies of the western United States. They take their name from their dependence on the evergreen sagebrush which provides them with both food and shelter. In spring, the cocks turn their attention to the setting-up of territories in readiness for the coming breeding season. Each territory covers only a few square yards of fairly open ground. It is used for neither nesting nor feeding but solely for displaying and mating. The territories are clustered together and as many as 400 cocks may establish themselves in one restricted area, known as a strutting ground, which is used year after year.

From February to May, the strutting ground is the scene of intense activity. The cocks arrive just before sunrise and assume the 'strutting posture' to indicate ownership of the territory. The pointed tail feathers are raised and spread into a fan. The white feathers and black plumes of the neck are erected into a ruff larger and more elaborate than anything Elizabethan courtiers could devise. At the same time the wings are lowered and yellow combs above the eye are expanded. At intervals of a few seconds a larger pouch in the throat is inflated to display two yellowish-green patches of bare skin on the chest. As the pouch is deflated, a resonant booming sound is made. Every now and then, males on neighbouring territories stop strutting and display at each other, advancing to the border between the territories until they are standing side by side and looking over each other's shoulders. They jostle to and fro and occasionally break into a fight, slapping each other with their wings.

The object of these displays is to decide the ownership of the territories and to space the sage grouse cocks over the strutting ground. When all are settled, the hens visit the ground, coming in first at sunrise and returning again in the evening. They do not visit all the cocks but gather in dense groups at 'mating centres'. Here they mate with those cocks whose territories overlap the perimeter of the mating centre, then depart to nest and raise their young on their own.

Mating is the cock sage grouse's sole contribution to the breeding cycle and he has been forced to evolve the elaborate plumage and displays of the strutting ground. But his chances of mating have been limited by the behaviour of the hens. As a result of the hens gathering at the mating centres most of the mating is carried out by the two or three adjacent cocks. Over the population as a whole, five to ten per cent of the cocks are responsible for mating with seventy-five per cent of the hens. The successful cocks are the experienced birds who have survived the rigours of life for several years and are, presumably, the 'fittest' males for producing offspring and ensuring the species' continued existence.

Polygyny, as the mating of one male with several females is properly called, is common in the bird world, particularly among the pheasant and grouse family. The male, of which the peacock is a familiar example, has brilliant plumage for displaying to females and other males. The female's drab plumage is a camouflage when she is sitting on the eggs. Communal display grounds are often used as a meeting place for males and females. They are called leks for black grouse, booming grounds for prairie chickens, hooting grounds for sharp-tailed grouse, hills for ruffs and arenas for birds of paradise. Wherever a close watch has been kept at meeting places, it has been found that a small proportion of males perform the majority of acts of mating. At a prairie chicken booming ground in Kansas one dominant cock alone mated with seventy per cent of the hens.

Two questions have not been properly answered yet. It is not known how some cocks become dominant over others. Young sage grouse cocks hold territories round the edge of the strutting ground. As birds stationed nearer the centre die, these young cocks take over their territories until, over a period of time, a few finish as old, experienced birds holding territories at the mating centres. But how the contestants for a vacant territory decide the winner is a mystery. Similarly, it is not known how the hens recognize the dominant cocks. At the mating centre in a strutting ground the sage grouse hens have no choice. They have access to only the dominant cocks, but reeves, the female ruffs, walk around the 'hill' and have access to any male. Yet they choose the dominant ruff and queue in front of him to be mated.

Where to nest
After birds have paired up, their next task is to find a suitable place to build a nest. Some birds have very strict requirements for siting the nest. There is

an Argentinian species of swallow which nests only in the abandoned nest holes of woodhewers (one of the woodcreeper family of the Americas) which are excavated in the sides of vizcacha burrows. So the distribution of the swallow is limited to places where another bird, the woodhewer, and a rodent, the vizcacha, live side by side. On the other hand, the emperor penguin has no nesting requirements. It carries its single egg on its feet and covers it with a fold of skin.

How a bird decides where to build its nest is not always clear. Pairs of chaffinches can be seen flying from tree to tree in search of a suitable site. The female takes the lead while the male follows, singing. The female alights on a branch and hops down it to the nearest junction. If this is a deep fork, she hops round and round, examining it. The process is repeated at several forks and eventually one is chosen and nest building starts. Presumably the chaffinch has examined various features of the fork and its surroundings and has decided that this particular one is right, that it has the correct shape, the right height above ground, is well concealed and so on.

What precisely confers suitability on a nest site can be determined by experiments and observation. Laughing gulls that nest regularly on a low island on the New Jersey coast were found to prefer areas just above the high water mark, where the grass is over 2ft high. As the gulls spend a month on the island before they build their nests they must learn which areas are not flooded by high tides. The grass may help conceal the nests from predators or reduce arguments between neighbours as, if a patch of grass is mown, the gulls will not nest on it.

Weaving nests

A bird's nest is a marvellous construction, particularly when we remember that the bill and feet of a bird have nothing like the dexterity of our hands. The nest of a blackbird or chaffinch looks as if it has been woven from dried grass, yet weaving seems an impossible technique for a bird to master. The process is more akin to the manufacture of felt because the bird twirls around on the nest foundations, pushing with its legs and working with the breast to mat the material into a compact mass. Weaving is, however, not an impossible task for a bird. The aptly named weaverbirds of Africa make elaborately woven nests from strips of leaves. The form of the nest varies between the 96 species. The red-vented weaver gives its nest a 2ft long entrance tube which collapses when a marauding snake attempts to enter and scores of pairs of sociable weavers live up to their name by constructing a communal nest that looks like a haystack perched in a tree. Beneath a single roof, each pair makes its own nest chamber.

Despite the variety of weaver nests, there is a basic pattern used in their construction. Nest building is carried out normally by the male, who starts by weaving a vertical ring of grass. Standing on the base of the ring, he continues weaving to make a roof which is extended backwards to make a nest

The emperor penguin is the largest of the family. The males incubate the eggs in the bitter Antarctic winters, crowding together in large numbers to conserve body heat. During the two month period of courtship and incubation, the males fast. The females hunt for their main food, squid, which they bring back in their crops at about the time the chicks are hatching. The females then take over the parental duties while the males seek food for themselves, but after about two weeks the now well-fed males return and both parents feed the chicks

chamber. Then, still in the same position, the weaver proceeds to extend the chamber forwards, adding material to the edge and leading the roof downward to make a hollow ball with a single round entrance near the base.

The strips of vegetation used in nest-building are taken from the long leaves of banana plants, grasses or palm trees. The weaver perches on the base of the leaf, bites through one edge and tears it back. Then, holding the torn end firmly in its bill, it flies away, tearing the rest of the strip loose. The strip of leaf is woven into the rim of the nest by tucking a loose end into the existing weaving and reaching behind to pull it through. This end is secured with a hitch and the free end is successively woven through adjoining parts of the fabric.

The completed nest is a compact firm ball but the first nest that a weaver builds is a loose tangle of loops and strands. Nest-building is instinctive in that a weaver, hand-reared in isolation from its fellows, can still build a nest but its technique needs to be improved by practice. It has to learn which materials are suitably flexible and it must learn the correct sequence of weaving actions. A young weaver wastes time by pushing strips in and then pulling them out before they are secured. It also discovers by experience that the strip has to be pushed in far enough to prevent it falling out while the bird is reaching around to hitch it and that the strip has to be held down by the feet while the bill grip is shifted. Until these important preliminaries are learnt, most strips end up on the ground under the nest.

Communal families

According to the book of *Genesis*, the animals went into the ark 'two by two'. We accept the pairing of animals as natural, perhaps because it accords with our own ideas of monogamous marriage. We see most of the common countryside birds forming pairs in the spring and rearing families in partnership. The same holds for seabirds but, as ornithologists investigate the habits of a wider variety of birds, they have found avian equivalents of the human communes which are being set up as a reaction against monogamy. Many birds share defence of the territory and care of the young. The group often consists of an adult pair with two or three young birds as helpers, but larger flocks also occur.

The Mexican jay lives in flocks of up to a dozen or so individuals. A flock consists of adult birds, with more males than females, and a number of yearlings which have stayed with their parents. They unite to defend the territory against other jays and to mob predators. Each adult female builds a nest and incubates her own eggs. During the incubation period she rarely leaves the nest and is fed by both her mate and the helpers—the surplus males and yearlings. When the chicks hatch out, they are fed by all the members of the flock, although the breeding females rarely feed chicks other than their own. The helpers also seem to have a favourite nest where they bring the bulk of the food, but when the young jays leave the nest, the helpers

Weaver birds are a scourge of the African savannah, flocking in millions and feeding voraciously on seeds. Their nests, which differ in shape from species to species, are miracles of construction, woven and knotted with feet and beak into dry, hollow homes, safe from predators and with their entrance on the underside, to prevent them being swamped by the rains

and even the breeding females feed each and any fledgling without favour.

Mimicking for survival

Talking birds are so familiar that they need no explanation. Parrots, mynahs and budgerigars are pets whose strange ability to imitate human speech is a constant source of amusement. But why do these birds, and several others, mimic us? The answer must come from a search for mimicking birds living in the wild. The mockingbird, whose scientific name *Mimus polyglottos* means 'many-tongued mimic', has a burbling, continually changing song. Up to ten per cent of the song is made up of mimicked sounds, including human voices and mechanical sounds. Starlings and jays sometimes bring calls of other birds into their repertoire but the Indian hill mynah, one of the best 'talkers', has never been heard to imitate other birds in the wild. This conclusion was reached after a two-year study of mynahs in the forests of Assam. However, it was found that young mynahs learn their songs by listening to other mynahs so that, if taken from the nest and hand-reared, they pick up human speech instead of mynah songs. Imitation is important for the development of singing ability in many birds. Chaffinches, for instance, cannot sing properly if reared in isolation. Part of the song is instinctive but the whole song develops only if neighbouring chaffinches are heard singing while the young chaffinch is still in the nest.

For the paradise widow-bird of Africa, learning which song to sing has acquired an unusual importance. Widow-birds live a life of deception. They are parasites like the cuckoo and lay their eggs in the nests of waxbills. A newly hatched cuckoo chick ejects its nest mates and receives the undivided attention of its foster parents, but the widow-bird chick is reared along with the waxbill chicks and adopts an elaborate disguise to prevent the waxbill parents from rejecting it. Waxbills of the family Estrildidae are unique among the perching birds for the way that they feed their chicks. They push their bills into the chick's throat and pump in predigested food. They are guided into the chick's mouth by its patterned lining. Each species of waxbill has its own pattern of black spots and lines on the palate and white or blue knobs around the edge of the mouth. The intruding widow-bird chick has, therefore, to mimic the mouth pattern of its nest mates, so each widow-bird species has mouth patterns that correspond with its host species.

For this ruse to work, it is essential that the female widow-bird lays her eggs in the nest of the correct waxbill. The paradise widow-bird is a special case because it is divided into seven distinct races, each of which lays its eggs in the nest of a particular waxbill. So chicks of each paradise widow-bird race must have the right mouth pattern if they are to live. They must be in the right nest and their mother must mate with a male of the same race. This is achieved by mimicry.

Male widow-birds learn the songs of their foster parents while in the nest. At the same time, female widow-birds are learning to recognize the same songs. When adult, they are attracted to male widow-birds that are mimicking the right waxbill song and, after mating, they seek a nest where the same song can be heard from the resident waxbill. The female widow-birds require two males in their lives. They mate with their own species but nest with the foster species, and both are recognized by the same song.

Voice keeps the family together

When the chicks of an Adelie penguin are three or four weeks old, they leave the nest and gather with the other chicks of the colony into a creche. Here they remain while both parents go off to collect food for them. On their return, they call out their own chicks from among the thousands in the creche. Other chicks may respond to the call but a parent penguin feeds only its own chick. To us, one penguin looks like any other but penguins have no difficulty recognizing each other. From the confused babble of harsh calls that makes bedlam out of a penguin colony, each chick can recognize the individual calls of its own parents.

Recognition of a mate is needed for harmonious proceedings of courtship and incubation. Herring gulls will awaken when hearing the call of their mates and gannets are alerted to their mate's arrival at the nest before they come into sight. The identification of the chick is not so essential and depends on the situation. Great skuas sometimes adopt chicks that have wandered in from neighbouring territories, but black-browed albatrosses nest in dense colonies on cliff tops. The chicks perch on drumshaped nests of mud set close together but the parent albatross lands at the edge of the colony and walks unerringly past scores of chicks to the correct nest. If chicks are swapped around, it will go to the same nest and feed the wrong chick. It recognizes the nest but not the chick. In nature, the right nest always contains the right chick and the system works because chicks never move from the nest until it is time to fly. The wandering albatross nests on flat ground behind the black-browed albatross colonies. Its nests are well-spaced and, when nearly ready to fly, the chicks often leave the nest and walk about, flapping their wings and practising take-off. The adults recognize their chicks and search for them if the nest is empty.

Knowing the identity of the chick is very important to guillemots which breed on narrow ledges on sheer cliffs. They pack together like commuting city workers on a railway platform and, as they make no nest, each guillemot has to recognize its own egg and chick. The egg is recognized by sight but sound is important for both chicks and adults to recognize each other. The chick starts to reply to the parent before it has even hatched and learns to distinguish the calls from those of other adults.

The critical period in a young guillemot's life comes when it is two to three weeks old. Before its wing feathers are fully developed, it flutters down from the cliff ledge and swims out to sea with one of its parents. Hordes of young guillemots take the plunge each evening and the water below the cliff becomes dotted with birds. The parents sort out their own chicks from the mob by identifying their voices. Each chick has a special call which the parents learn a few days after it has hatched.

Sharing out the food

Among the marvels of the natural world are the huge gatherings of birds that present a riot of colour and movement. Rocky islands of the Antarctic form the breeding grounds of penguins where they may gather in tens of thousands. In the northern hemisphere penguin colonies are replaced by cliff-ledges swarming with guillemots and razorbills and warrens of puffin burrows, where the sky is clouded by the birds as they fly to and from their nests. In between, the inland waters of the tropics provide food for a spectacular variety of birds. Lake Nakuru, the best known of the African Rift Valley lakes, supports thousands of fish-eating birds. There are a thousand pelicans that commute in from distant nesting grounds, as many cormorants and darters, as well as storks, spoonbills, gulls, terns, a variety of herons and a few skimmers and ospreys: an incredible profusion of bird life feeding on one fish—*Tilapia grahami* which was introduced a few years ago to control malaria-carrying mosquitoes.

A thoughtful observer of these masses of birds will ask how the multitudes can be fed and how it is that the numerous or powerful species do not oust their fewer and weaker neighbours. It would seem that they must compete for a limited, albeit vast, supply of food but their co-existence lies in their ability carefully to avoid competing with one another for the same food. The ways by which they achieve this is being realized by careful studies of how and where each species feeds and what it is eating.

The diets of many seabirds have been quite easy to study because they can be persuaded to regurgitate the food they bring back to feed their chicks. Analysis of this food has yielded some subtle, and sometimes surprising, differences in diet between birds living in the same place.

On the islands of Skomer and Skokholm, off the coast of Wales, there are colonies of three species of auk; the puffin, the razorbill and the guillemot. They feed mainly on sand eels, sprats and small herring but, even when all three auks are catching the same kind of fish, they do not compete directly. The guillemot catches the largest fishes and carries them home singly for its chick. At the other extreme, the puffin catches the smallest fishes and can carry up to thirty at a time. The razorbill strikes the medium and carries a few middling fishes. The difference in diet is related to the size of the bill. The guillemot has a long pointed bill, the puffin has the familiar broad, conical bill with its colourful stripes, and the razorbill has a bill of intermediate shape.

Difference in bill size as a means of avoiding competition is taken to extremes by the Adelie penguin of the Antarctic. In the Ross Sea, along the coast where Amundsen, Scott and Shackleton based their expeditions, the Adelie penguin feeds mainly on shrimp-like crustaceans called krill. Krill are also the main food of the great whales. Male penguins are larger overall than the females and the difference is greatest in the size of the bill. The result is that males eat bigger krill and it is possible that the slight difference in diet based on bill size may reduce competition for food between the sexes.

In the Galapagos Islands that straddle the equator, there are three kinds of storm petrel. These are the smallest of the petrel group, which contains birds up to the size of the albatrosses. The three species avoid competition for food by their feeding habits. They catch small fishes, crustaceans and squid at the surface, but the Madeiran storm petrel specializes in fish whereas the Galapagos storm petrel eats some crustaceans and catches smaller fish than the Madeiran storm petrel. Of greater importance, the Madeiran storm petrel feeds by day while the Galapagos storm petrel feeds by night. The third species, Elliot's storm petrel, avoids competition by feeding in inshore waters, leaving the open sea to the other two.

There are also three species of boobies, relatives of the gannet, living on the Galapagos Islands. They avoid coming into conflict by choosing separate feeding grounds. The blue-footed booby feeds inshore, close to the islands, the red-footed booby feeds well out to sea and the masked booby feeds between the others. The feeding grounds have an important bearing on the boobies' breeding habits. The short distance between feeding grounds and nest allows the blue-footed booby to rear up to three chicks. It has no trouble bringing enough food for them but the masked booby lays two eggs and rears only one chick. The red-footed booby, hampered by its lengthy commuting time to distant waters, lays only one egg.

Oystercatchers learn to feed

Oystercatcher is rather a misleading name for a bird that eats a wide variety of seashore life, such as cockles, crabs, winkles, limpets and lugworms. Musselpecker is an old country name for the oystercatcher and the large beds of mussels which are exposed at low tide are a favourite feeding place. The oystercatcher has two methods of dealing with a mussel. If the mussel is covered by shallow water, its two shells will be agape so that it can draw into its body the current of water that is essential for feeding and respiration. This makes is very vulnerable to oystercatcher attack. The bird's long, awl-shaped bill stabs downwards and severs the stout adductor muscle that closes the shells. The mussel is now defenceless and the oystercatcher can prise the shells apart with its bill and nibble the flesh at leisure. If the mussel is exposed by the falling tide it stays tightly closed and the oystercatcher adopts a different technique. It tears the mussel from its anchorage and carries it to a patch of sand. There, the mussel is placed with the hinged side uppermost and the oystercatcher hammers it with its bill. About five blows are needed to drive a hole through a mussel shell, and if the mussel falls over under the impact, it is carried to a firmer patch where it will sit firmly. Again, the bill is used to cut the adductor muscle and to prise the shells apart.

Examination of the shells left by oystercatchers shows that they regularly attack the broad, hinged edge of the mussel and tests have proved that this is the easiest place to force an entry. Cockles are attacked in the same way but the shell is hammered anywhere as it is weaker than a mussel shell. Crabs receive a different treatment. They are thrown on their backs, despatched by a stab through the brain and their shells levered off.

Surprisingly, an individual oystercatcher is not adept at each of these methods of killing prey. It is a specialist; an oystercatcher that stabs mussels does not hammer them and *vice versa*. This means that a large flock of oystercatchers feeding on a shore is divided into groups. The 'hammerers', for instance, feed above the tide mark while the 'stabbers' work in shallow water beyond the water's edge. Moreover, young oystercatchers follow their parents' habits. Not long after they have hatched, the chicks are lured from the nest towards the shore by the parents calling and presenting food just out of reach. When about three weeks old, the chicks start to feed themselves. At first, they play with empty shells, picking up pieces of flesh left in them and practising the 'scissoring' movements used by the adults for tearing flesh off the shell. Later, the parents bring opened shellfish and the chicks feed themselves. Eventually, they learn both to recognize and to open their own shellfish. So ingrained is this reliance on one feeding habit that a mussel-eating chick is actually scared of crabs, an otherwise common food of the oystercatcher species. Furthermore, a stabbing mussel-eater will mate with a stabbing mussel-eater and not with a hammering mussel-eater or a crab-eater.

Stone-throwing vultures

At one time biologists thought that tool-using was the character that set Man apart from all other animals but, as more animals have been studied, a select few have been found using tools. It is true that animals rarely make their tools and usually use convenient objects which come to hand, and that they do not display much finesse in their tool-using; but nonetheless they qualify as true tool-users. Tool-using involves the use of an object to increase the efficiency of an action. Lifting a boulder is carried out more efficiently with a crowbar.

The Egyptian vulture of Africa uses stones as tools to break open ostrich eggs. This is not a common source of food for vultures but nests are occasionally abandoned and the eggs are there for the taking. Vultures cannot smash through the hard shells of ostrich eggs with their bills but the Egyptian vulture gets round this drawback by throwing stones. It will pick a stone, carry it towards the egg, then stretching up to its full height with bill pointing skyward, it flings the stone down at the egg. A dozen strikes may be needed to fracture the eggshell and, as half the throws miss the egg, breaking an ostrich egg takes some effort. But it is worthwhile because a $3\frac{1}{2}$lb egg is a good meal for a raven-sized vulture.

Unlike many other vultures, the Egyptian vulture has feathers on its neck and head and a narrow bill. While other species may feed by plucking meat from deep inside carcasses and are bald to avoid contamination with the rotting food, Egyptian vultures are content to pick up the scraps from the periphery. They are among the comparatively few tool-using animals, and here one is shown about to smash an ostrich egg with a stone. They are found in southern Europe, North Africa and throughout the Middle East and India

Butcher birds

The shrikes are perching birds related to the familiar songbirds of garden and country, but they have adopted the habits of birds of prey. They kill small animals, such as beetles and grasshoppers, frogs, lizards, small birds and mammals, and will even attack animals larger than themselves. After a chase or a quick swoop, the prey is knocked to the ground and despatched with a blow of the bill. Shrikes have unusually large powerful heads and the bill is hooked and 'toothed' like the bill of a falcon. The 'teeth', on the cutting edge of the upper mandible, may help to sever the spinal cord as the shrike slices at the neck of its prey.

Such able hunters are the shrikes that they have been kept for hawking. King Louis XIII of France kept shrikes in the royal mews and they have long been favourites in oriental courts. The principal difference in the hunting techniques of true birds of prey and shrikes is that shrikes lack talons and strong feet. They cannot hold down large prey, such as mice or birds, while flesh is being ripped off with the bill. Instead, the shrikes secure their prey by impaling it on a thorn or wedging it in the fork of a branch. Barbed wire is used as a substitute for thorns on farmland. The prey is dragged against the thorn or fork until it catches and holds firmly in place while chunks are torn away.

In the course of a breeding season the territory of a shrike becomes festooned with the sad, dessicated and dismembered carcasses of small animals hooked through the skin. These macabre relics gave rise to the name of butcherbird and to the idea that the shrike kills for fun. Although more animals are killed than is necessary at the time, the caches or larders are used as reserves for feeding the young or when food is scarce. Insects, which can be eaten without impaling or wedging, are often stored, and loggerhead shrikes of North America impale lizards, returning to them in cold weather when live lizards have gone to ground.

Finding the way

The ability of birds to find their way to winter feeding grounds and then migrate back to the same place, often to occupy the previous year's nest, has long been a wonder of the natural world. One of the more incredible feats is that of young cuckoos which set off for their winter home one month or more after their parents. Clearly they cannot learn in which direction to fly or when they have arrived at the right place. The ability to navigate must be inborn.

A young cuckoo, or any other migrant bird, carries in its head all the paraphernalia of a human navigator. It has a sextant, a chronometer, the charts and sailing directions that tell it where to go and, perhaps, a magnetic compass. No one has actually found the equivalent of these instruments in a bird's brain but we can infer that such mechanisms must exist from experiments on captive birds and by observation of wild migrants. The principle of navigation is to establish one's position, compare this with the position of the destination and then calculate the necessary course to link the two. A bird establishes its position by the sun. Using the equivalent of a sextant it observes the movement of the sun along its arc for a short time and extrapolates the line of the arc to give the elevation of the sun at noon. This gives the latitude. To calculate longitude, the bird uses its 'chronometer' to compare the difference in time between noon at its present position and noon at its destination, the latter being part of the bird's inherited 'navigation instructions'. The position of the destination is also inherited and the young cuckoo sets off on the appropriate compass course.

In practice the migrant meets bad weather and contrary winds which will upset its navigation. Birds prefer to start their journey on a fine day when they can take accurate fixes of the sun's position and when the wind will not immediately blow them off course. Strong winds met during the flight may throw the migrants many miles from their route but, providing they can check their position, they can alter course accordingly. Overcast weather stops the birds from taking-off, but if the sky clouds over while they are in flight they will keep going. Like the human navigator, they can steer by 'dead reckoning'. The human navigator uses a compass to maintain his course and hopes that the sun will eventually break through and allow his position to be fixed. It is possible that migrant birds do the same thing. Some birds, at least, have a magnetic sense that enables them to maintain a correct heading even in complete darkness.

Fuel for migration

Life for a bird is not easy. Predators, food shortages, bad weather and disease are constant hazards, but migration is an additional hazard that must be faced twice a year. A successful journey depends largely on the bird having sufficient fuel food reserves for the strenuous flight. For some migrants the situation is eased by a leisurely migration with frequent stops for feeding. Some chaffinches, for instance, migrate from Norway to Britain across the North Sea but others make a roundabout trip down the European coast. They pass through Denmark and Holland to the French coast where they make the short sea-crossing into southern England and spread through the British Isles. This itinerary is three times longer than the direct sea crossing but the risk of becoming exhausted and crashing into the sea is much less.

If there is an advantage in taking the long way round, it cannot be sufficient to prevent many migrants risking the difficult direct route. The ruby-throated hummingbird crosses the Gulf of Mexico rather than flying around its shores, and millions of small birds migrate between Europe and Africa by crossing both the Mediterranean Sea and the Sahara desert. The combined crossing is more than 1000 miles and apart from a few islands, oases and the coastal strip there is nowhere to land and feed.

The Sahara crossing is made by swallows, sand martins, wagtails and many

Birds such as warblers will migrate non-stop in the spring from the heat of Africa, across the Sahara and the Mediterranean in one exhausting flight to gain the more temperate climate of Europe. The journey may take 60hrs, and during this time 30% of the body weight or 60% of the fat reserves will have been used as fuel for the flight

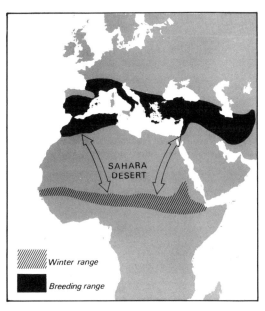

SAHARA
DESERT

Winter range

Breeding range

kinds of warblers. These are small birds weighing in the order of 10-20 grams ($\frac{1}{3}$-$\frac{2}{3}$oz:). It is incredible that they, or the diminutive ruby-throated humming-bird, are able to make such an arduous trip. Many perish, to be sure, but for the species to continue many more survive. The secret of their survival lies in the fact that the distance they can fly is proportional not to the total amount of fuel carried but to the ratio of fuel and body weight. A small bird does not need a huge reserve of food but one proportional to its size.

The most efficient fuel for an animal is fat. Each gram burnt in the body provides over twice as much energy as a gram of protein or carbohydrate. At the same time it yields a gram of water, which is a great asset on a desert crossing. Before migrating birds put on a large amount of fat. Some almost double in weight and the resulting fifty per cent ratio of fuel to body weight gives them a very good range. Warblers returning from Africa pause in the savannahs and woodlands and gorge themselves with food. They even abandon their usual insect diet and feed on berries. Fat is laid down at a rate of up to 1 gram per day and as soon as it is fully fattened, the migrant sets out. Flying non-stop, it uses about 0.5 per cent of its body weight every hour. 4-5 grams of fat on a 12-gram bird is quite sufficient for a non-stop journey across the Sahara and the Mediterranean. Even allowing for some headwinds the journey would take no more than sixty hours. In this time thirty per cent of the body weight, or sixty per cent of the fat reserves, will have been used.

Once we realize that it is the proportion of fuel carried to body weight that is important, there is nothing remarkable about long flights by small birds. Large birds cannot make such long non-stop flights because it is physically impossible for them to take off with a large fat reserve. The most efficient migrants are the medium sized birds that can carry large reserves and are strong enough to withstand adverse weather. The Eskimo curlew flies from Alaska to Argentina. First, it make a detour to Labrador and Newfoundland where it stuffs itself with crowberries before flying down the east coast of the Americas. At one time Eskimo curlews were shot in their thousands as they passed through the New England stages. As they hit the ground their bodies split open to reveal a thick layer of fat.

Ranging from the Tropic of Cancer to the Arctic Circle, the golden eagle is nowhere common. Its main food is birds and small mammals, though it will occasionally eat carrion. Its nest, or eyrie, is built of sticks roughly piled together high up on mountain ledges

Above A pair of western grebes in the Sand Lake Wildfowl Refuge, South Dakota, skitter across the surface of the lake during an elaborate courtship display. Grebes specialize in diving from the surface for their food, mostly fishes and shellfish

Right The ostrich is the largest living and the fastest running bird. It is found through much of Africa. 8ft tall, it can outstrip most predators, sprinting at speeds of up to 40mph. Largely a bird of the desert, the ostrich can drink salty water, getting rid of excess salt through a nasal gland. These two cocks, their feathers ruffed out, are disputing territory

Left Migrating blue and snow geese fly as much as 2000 miles non-stop in strict V-formation during their twice yearly migrations from the Arctic tundra in September to warmer wintering grounds in southern North America. They return in May

Above Many birds change the colour of their plumage as the seasons change. The willow ptarmigan demonstrates how well the plumage changes suit the background through an Alaskan year

Birds get their food in many ways. The puffin (*left*) feeds mainly on sand-eels (*Ammodytes*), a delicacy even for humans in many countries. The American anhinga or snake bird actually spears its fish with its straight sharp bill. The great white pelican chick takes its food from the back of its parent's throat after it has been regurgitated. Pelicans hunt cooperatively, ladling up the surrounded fishes into their great throat pouches

Amphibians and reptiles

by Dr Ulrich Gruber

Introduction to amphibians and reptiles

The first vertebrates to emerge from the sea on to the land, some time between the Devonian and the Carboniferous periods, about 350 million years ago, were the amphibians. They are thus the oldest of the land vertebrates. Yet to this day they have still not been able to free themselves completely from the watery element, for most species still pass the first stages of their lives in water, as larvae with gills for breathing or as tadpoles with tails, until after a fundamental change (metamorphosis) they clamber on to the land as little newts, baby toads or tiny frogs. Of course there are exceptions to this among all the groups of amphibians, among the caecilians, the newts and salamanders, and the frogs and toads. Notable exceptions include mountain species such as the alpine salamanders, which because of the bitter climate of their habitat complete the embryonic development stage within the maternal body and bear their young living. Other species, such as many of the Antilles frogs in tropical America, need only the little drops of water in bromeliad plants to lay their eggs in. Others again, like the female marsupial frogs or the bowl-backed tree toads in Brazil, keep their eggs safe in a pocket of skin on their back, where they find protection and the essential moist surroundings, until the first tadpole stage. Even out of water, amphibians require high humidity in the surrounding air. Their skin is kept permanently moist by means of mucous glands distributed over the whole of their body, and they have always to lie low to avoid being dried out. There are just a few species which have horny skins that protect them better against drying—the green toad in Europe and Asia is one—and these can live even in desert regions. Many amphibians have poison glands in their skin, as toads have, for instance, near their ear at the back of the head, and the poison from these can cause an irritating inflammation of the mucous membranes of an aggressor.

Among their special characteristics, most amphibians have four fingers and five toes. Respiration and blood circulation are also distinctive features of this group of animals. We know of examples of throat breathing, gill breathing, lung breathing and breathing through the skin; among these methods skin breathing is of particular importance. An amphibian heart has two auricles, unequal in size and only fully separated in the frogs, and a single ventricle in which arterial blood and venous blood mix freely. The lymph circulation is perfectly developed. Among the senses, those of touch and sight are the most important. In general, the function of the eyes is chiefly

Poison arrow frogs have no need to disguise their presence. They manufacture a highly poisonous toxin, with which South American Indians coat the tips of their arrows and spears

limited to observing movement; nearly all amphibians see and snap up their prey only when it moves.

An outstanding feature of amphibians is their voices. Who has not heard, on mild summer nights, the concert of marsh frogs and edible frogs in ponds and pools, or the deafening, hammering 'ke-ke-ke' of the European tree-frogs? Often the call is amplified by means of bladders which are inflated to give resonance, as for example in the edible frogs, or which are blown out as strange sacs from the throat, as with the European tree-frog and the natterjack.

All amphibians—salamanders, newts, frogs and toads—are dependent on the temperature of their surroundings for their degree of activity. They are, as we say, 'cold-blooded'. If the surrounding temperature falls below a certain minimum the animals trim their activities to the bare essentials and begin a 'hibernation', and it is only this hibernation that makes it possible for them to maintain a continued existence in the temperate zones of Europe, Asia and North America, or in the middle and upper levels of the great mountain ranges, the Alps, the Himalayas, the Andes. In the spring, when the hibernation is over, both sexes of the amphibians move to the spawning sites and mate there. Among tailed amphibians, such as newts and sala-manders, fertilization is as a rule internal, the female receiving the spermato-phores deposited by the male with the lips of her cloaca. With frogs, on the other hand, the eggs are fertilized outside the body; the male, clinging fast to the female's back, sprays his semen over them the moment they emerge from the female's cloaca. The eggs are often laid, as they are for instance by our common toads and common brown frogs, in huge numbers, in jelly-like strings or clusters in ponds and pools. Many of the frog species take great care of their brood. The male of the Western European midwife toad, for example, drags the eggs around with him wrapped round his hind legs and spreads them on the surface of the water only the moment before the tadpoles hatch out. We have already mentioned the pouch on the back of the marsupial frog. The grey tree-frog of South America makes a nest of froth on twigs high above the water, from which the tadpoles drop into the water when they hatch out.

The class Amphibia is divided nowadays into three orders: the caecilians (Caecilia), the tailed amphibians (Urodela) and the anuran amphibians (Anura). The caecilians, which look like outsize earthworms live only in the sub-tropics and tropics. The tailed amphibians look lizard like and mostly live in the northern hemisphere. This order of amphibians includes some that are quite like the European mountain newt and the Mexican pygmy sala-mander, only about $1\frac{1}{2}$in. long, and others that are astonishingly big, like the giant salamanders of Japan and China, which can reach a total length of some five feet. And finally the anuran amphibians, with their tailless bodies often equipped with long hind legs for jumping, embrace a large number of sub-orders and families, including the true frogs, the toads and the tree-frogs.

Reptiles, like amphibians, are described as cold-blooded animals; that is to say, the level of their activity is controlled by the temperature of their surroundings. Their skin is not generally naked and damp like that of the amphibians, but dry and covered with horny scales or shells. With a few exceptions—the sea-snakes, turtles and sea-lizards in the sea, and the crocodiles and marsh-turtles in fresh water—they are true land animals. Their embryonic development no longer goes through a larval stage in water, but is completed in eggs rich in yolk, which are either laid on land or remain within the mother's body until they hatch.

The reptiles that exist today are no more than a remnant of a race of countless widely varied species that flourished in the middle ages of the earth from the Triassic to the Cretaceous periods, some 200 million to 100 million years ago. One characteristic that has come down to them from prehistoric times, when the climate was generally warmer, is their great love for the warmth of the sun. Their hunger for the sun is almost insatiable. If the surrounding temperature falls below a certain level, reptiles, like amphibians, begin to hibernate, and it is this hibernation, with a greatly slowed down metabolism, and bodily functions reduced to the minimum, that makes it possible for many species to exist in temperate zones, in mountains and in a few cases actually within the polar circle. The European adders and viviparous lizards are examples. However, the great majority of reptiles do live in the sub-tropics and tropics.

Their dry, horny skin, with its covering of scales, protects the reptiles' bodies from damage and especially against dehydration. Consequently they are to a great extent independent of water and can even live in deserts. The common skink, 'swimming' like a fish through the loose desert sand, the sidewinder gliding sideways over the ground like a ghost, or the herbivorous spiny-tailed lizards are examples. These animals get all the moisture they need from their food. Lizards and snakes generally find their growth inhibited by their hard skins, and so from time to time these skins have to be sloughed. Snakes shed their skins entirely, like a glove pulled off a hand inside out, and we may sometimes come across these transparent discarded skins. Lizards cast their skins bit by bit, in little pieces. Most geckos finish by eating their skins after sloughing them, some completely, some only partly.

The bodies of turtles and crocodiles are completely insulated and specially protected from exterior damage. Turtles and tortoises have bony armour covered with horny shells and often also with a leathery skin, protecting both back and belly. Crocodiles are enclosed in an armoured integument formed of large, strong, often bony plates.

Most reptiles are four-footed, usually with five fingers and five toes. But many forms, such as the slow-worms, the unique worm lizards (Amphisbaenidae) and the large sub-order of the snakes, have completely lost their limbs. These lengthy, footless creatures move with a winding motion; snakes also use their ribs, which are hinged to their spine, to help with their movement. This mode of progression, in for example the Gaboon viper of tropical Africa, looks like the crawling of a caterpillar or perhaps a children's sack-race.

Reptiles have well developed lungs. Their heart has one double auricle and an incompletely divided ventricle in which venous blood and arterial blood are still to some extent mixed. Of the organs of sense, the eyes and the sense of smell are well developed. Lizards and snakes smell with their tongues as well as with their nostrils. As their forked, moist tongues flicker outside their mouths they trap particles of the substance stimulating the smell from the air and transfer them with the tip of the tongue inside the mouth to a double scent-cavity on the roof of the mouth (Jacobson's organ). This cavity is equipped with a scent-epithelium and acts as an organ of smell. Lizards have good hearing, but snakes are deaf. In compensation, they are extremely sensitive to movement. A peculiar feature of the pit-vipers, such as the rattlesnakes or the South American lance-head snakes, is a pit-shaped organ sensitive to warmth, located between the eyes and the nostrils, with the aid of which they can locate warm-blooded prey like mice and rats even in pitch darkness.

Fertilization is internal in all reptiles. Most species lay eggs; many kinds are viviparous—the European viviparous lizard, for example, various chameleons and in the snake family most of the vipers (Viperidae). In the viviparous lizards and snakes the young emerge from the eggs immediately before or at the moment that the eggs are laid; the process is called ovoviviparity. The mating habits of lizards and snakes are interesting, mating being preceded by ritual 'love-play'.

The experts have divided reptiles into four orders, with six sub-orders and 48 families. The four orders comprise the tortoises and turtles, the crocodiles, the tuatara (of which only one family, with a single species, still exists, in New Zealand) and the true scaly reptiles, divided into the sub-orders of snakes and lizards. The long history of their development, going back to remote periods in the history of the world, the powerful urge to seek a warm environment and the consequent preference for a tropical or sub-tropical habitat, have quite simply turned the reptiles into a biological speciality, uniquely adapted to their environment and equipped with special features. We have only to think of the shells on the turtles, the crocodiles' armoured hide, the lack of feet in slow-worms and snakes, the chameleons' prehensile fingers or the poison fangs of the sea-snakes, cobras, vipers and pit-vipers. Their poisonous nature adds to most people's interest in snakes an element of the sensational, of the horrific and mysterious. Venomous snakes have poison glands in the upper jaw and two or more big poison fangs in the upper front row of teeth that are erected when the jaws are open; in sea-snakes and cobras these are fixed, but in vipers and pit-vipers they can be raised and lowered. When the snake bites, the fangs work like hypodermic syringes; there is a tube running through them ending at the point of the tooth in an outlet through which the poison is injected into the wound.

Salamanders that always remain as larvae

In the caves of Postojna in Yugoslavia lives the cave-olm (*Proteus anguinus*). What little information we have about the life of this species has come from observing animals in captivity; very little is known about how they live in the wild state. Only the physical conditions in the dark, subterranean dwellings of the cave-olm have been carefully recorded. The water temperature varies very slightly from winter to summer, with a minimum of about 10°C (50°F) and a maximum of 17°C (63°F); the water itself is medium-hard (pH-value 6.3) and exceptionally rich in oxygen. Cave-olms are adapted in many ways to existence in such surroundings. They have external gills all their lives, so that they seem to be permanent larvae, never able to develop into a hypothetical land form. Even under laboratory conditions it has never been found possible to induce metamorphosis—not even by the injection of thyroid hormones, and they reproduce in this permanent larval stage. Many other salamanders can do so, a peculiarity known as neoteny. The skin of the cave-olm appears milk-white; only the gills, with three bunched branches on either side, stand out in contrast as bright carmine where the blood flows strongly through them. However, if you gradually take the animals out into the light of day, they slowly take on a dark colour—a colour which changes back to the normal milk-white when the olms are put back into the dark. The rudimentary eyes are hidden beneath the skin, and the cave-olm is practically blind. The legs have only three toes on the forefeet and two on the hind feet. The olm lives mainly on small shellfish, but in captivity it also eats worms and small fishes. It has been possible to observe mating and egg-laying in the terrarium. There is a mating ceremony before the actual mating. The female lays up to sixty eggs and watches over them at the spawning site. But living young are also born, and it seems that most of the eggs laid while the female is carrying the young before birth serve primarily as food for the few embryos reaching full development.

The axolotl (*Ambystoma mexicanum*) belongs to the family of mole salamanders (Ambystomatidae), a family which is distributed from North America to Mexico. But free-living axolotls occur only in Lake Xochimilco, south-east of Mexico City. This salamander, too, generally reproduces while still in the larval stage, and keeps its gills all its life. However, if the water in which it lives gradually dries up, or if it is fed with thyroid extract in the laboratory, it can develop into a land form without gills. Then if the offspring of such fully developed axolotls are put back in their hereditary water environment, they again remain in the larval stage and reproduce in that stage. It has sometimes been possible in the aquarium to get albino specimens as well as the dark coloured form that occurs naturally.

The western sub-species of tiger salamander (*Ambystoma tigrinum*), found in the United States, can also reproduce in the larval stage, with fully developed gills. The cause of this phenomenon is thought to be a lack of iodine in the water and consequent inadequate production of thyroid hormones. The low production of hormones might also be attributable to the particularly low temperature of the water in which the creature lives.

Neoteny (reproduction in the larval stage) has also been observed in the Texas blind salamander (*Typhlomolge rathbuni*) and the Georgia blind salamander (*Haideotriton wallacei*). The Texas blind salamander grows to a length of $5\frac{1}{2}$in., has very thin legs and keeps its gills all its life. It is to be found only in wells and in the underground water in the mountains of mid-Texas. The Georgia blind salamander, about 3in. long, shows no trace of any eyes. It was first discovered in 1930, in a well nearly 200ft deep in Albany, Georgia, but it also lives in cave waters.

Spadefoot toads

The spadefoot toads of the genus *Scaphiopus* live in the desert regions of the United States and northern Mexico. These dull brown, warty toads, with their big eyes and the hard shovels on their feet, are exclusively night creatures. All day, and all through the long dry season, they lie buried in the loose, sandy soil. Only under cover of night, when dew falls on the ground, can they engage in their hunt for insects, other arthropods and worms. But the moment a heavy, warm rain falls on their home, hundreds and thousands of spadefoot toads suddenly appear, as if by magic, and there begins a breathless race between the reproduction process and the threat of death from the drying up of the waters. The spadefoots spawn in a single night; mating is already completed by the next morning. Two days later the tadpoles hatch and start on a frantic development. In great swarms, often including the young of several species, the tadpoles move through the water, stirring up the mud with their tails, and eating . . . eating. They stuff huge quantities of food into themselves to make themselves grow as quickly as possible. Sometimes—and it has not been learnt how this happens—many of the tadpoles' jaws, which are adapted for eating algae, develop into predators' jaws. They then turn from algae-eaters into cannibals, diving on their brothers and sisters and devouring them, to give themselves an even better chance of rapid growth. Or it can happen that predatory tadpoles of the plains spadefoot (*Scaphiopus bombifrons*) devour the algae-eating larvae of the southern spadefoot (*Scaphiopus couchi*) hatched in the same pool.

Finally the tadpoles turn into tiny, young toads—often at the very last minute, just as the waters finally dry up. Then whole swarms of young toads, the size of flies, break out on all sides. Within minutes such a swarm of young spadefoots, all metamorphosed at the same time, will have disappeared, seeking shelter from the murderous sun in some damp hiding-place. The landscape shimmers in the heat, bare and deserted, and nothing remains to tell of the ruthless battle for survival that has just been fought.

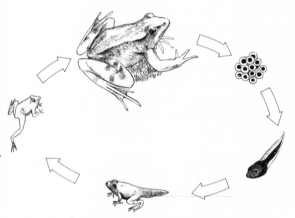

The life cycle of the frog shows all the typical amphibian characteristics. In the early spring the common frog emerges from hibernation (spent in a hole in the ground) and makes for water. The male develops a horny cushion on his first finger with which he can grip the female during mating. As the female lays the mass of jelly-like eggs in the water the male fertilizes them. The eggs quickly develop and the tadpoles hatch in about 2 weeks. At first they have external gills for breathing and no limbs. After a while the external gills disappear and are replaced by internal gills which themselves are eventually replaced by lungs. Hind legs appear before fore legs and the tail finally diminishes in size and is absorbed into the body of the now frog-like little frog. The whole tadpole stage takes 3–4 months. After this time the frogs leave the water and live in grass, feeding mainly on insects and small worms. As autumn comes they grow slow and lay down fat in a gland in the abdomen. Eventually they hibernate in a hole in the ground, living through the winter on their stored fat

Frogs with eggs on their backs

The two great stimuli in the animal kingdom are food and reproduction. This is especially true of frogs. In order to afford the maximum protection to their relatively small spawning clutches, many amphibious frogs have developed their own special ways of caring for the eggs and of the hatchlings while they are still young. The females of the marsupial frogs (genus *Gastrotheca*) in Brazil, for instance, have a pouch of skin on their back which serves as a receptacle for the eggs. Robert Mertens, the reptile authority, was the first to observe the egg-laying of the giant marsupial frog (*Gastrotheca ovifera*). The male, which is very much smaller, sits on the female's back during mating, holding on to her arms with his hands. Just before the eggs are laid the female straightens her hind legs so that she is standing bent forwards at an angle of 20° or 30°. Now, when the eggs emerge one by one from the female cloaca, they slip down to the front, under the male, as if they were on a slide. There they are fertilized, after which they go on into the pouch on the female's back, the male helping them on their way with his hinds legs. In this way about twenty eggs containing much yolk reach the frog's pouch, where the embryos also develop to the perfect frog stage. After a few weeks the female releases the fully metamorphosed young frogs from the pouch, holding the folds of skin open with the toes of her hind feet so that the young frogs can slip out. Since the giant marsupial frog lays comparatively few eggs, the development of the young can proceed in the pouch until metamorphosis is complete. In the common marsupial frog (*Gastrotheca marsupiata*), on the other hand, which has to get up to 200 eggs packed into its pouch in two shifts, the young have to be released while they are still tadpoles. They complete their metamorphosis in little puddles of rainwater or in water-filled funnels among the leaves of fleshy plants known as bromeliads.

There are other species of frogs that carry their eggs on their backs. The bowl-backed tree toad (*Fritziana goeldii*), for example, is a tiny frog that lives in Brazil. The female carries her twenty to thirty eggs in a bowl-like pouch on her back, covered with transparent skin. This frog too releases its tadpoles shortly after hatching in bromeliad 'funnels' full of water.

The females of the Surinam toads, members of the pipid family (*Pipidae*) of northern South America, such as the common Surinam toad (*Pipa pipa*) and the pygmy Surinam toad (*Pipa parva*), carry their eggs and their young until they are fully developed young toads in honeycomb-shaped depressions on their backs. Egg-laying among these toads requires a complicated ritual. When the eggs emerge from the female cloaca, the two partners, clasped firmly together, turn once round their lengthwise axis and in doing so make the eggs slip forward a little, between the female's back and the male's belly; there they are fertilized and sink into the skin of the mother's back. After three or four weeks the complete young toads are let out from the honeycomb-shaped holes into the water.

Some strange forms of turtle

The fringed turtle or matamata (*Chelus fimbriatus*), one of the family of snake-necked turtles (*Chelidae*), lives in tropical South America. Its flat head is extended forward to make a kind of flexible snout, its cheeks are drawn out like triangular sails of skin, and tufted flaps of skin hang down from its throat. On the top of the flat upper shell, three rough ridges run from front to back. This bizarre shape makes first-class camouflage for the matamata when it is lying in wait for its prey on the bottom of the water. As soon as a small fish swims by within range of its jaws, it suddenly opens its huge mouth with a loud, smacking sound. This quick movement creates an undertow, which drags the fish into its gullet in a fraction of a second. Before anyone has noticed what is happening, the matamata shuts its jaws again, gulps once or twice, then lies motionless again on the bottom.

The southern United States is the home of another turtle, the alligator snapping turtle (*Macroclemys temminckii*), one of the family of loggerheads (*Chalydridae*). This turtle is a specialist in hunting from ambush. With a shell up to 2ft 6in. long, and weighing up to 220lbs, it is one of the biggest species of freshwater turtles. Its shell has even higher ridges on it than those on the matamata, and its jaws make a hooked, vulturine beak. It sits motionless on the bottom of the water with its jaws wide open, and 'fishes' for its prey. There is a pink extension on its tongue, like a worm, that darts hither and thither. As soon as a fish sees this extension it swims up to it, taking it for a worm, and tries to eat it. Like a flash the hard jaws snap together, and the fish is swallowed. While the matamata sucks its prey in, the alligator snapping turtle just waits until its food swims into its mouth. Unfortunately the number of these turtles is becoming seriously reduced.

It is one of the wonders of nature that the biggest tortoises are those that live on tiny, remote islands. Biggest of all is the Seychelles giant tortoise (*Testudo gigantea*) in the Seychelles, north of Madagascar, and next to it comes the Galapagos giant tortoise (*Testudo elephantopus*) in the Galapagos Islands, which lie far out in the Pacific. The shell of this tortoise measures over 3ft and it can weigh more than 500lbs. The shortage of water in the Galapagos Islands has given these giant animals a remarkable way of life. They rest, basking in the sun, in the dry warm lava soil of the lowland parts of the islands, while their drinking water and food come from the fresh-water springs and rich plant life of the uplands. For countless ages the tortoises have been steadily marching to and fro between the two areas, making broad, smooth footpaths in the rough lava soil. Once man had discovered it, the Galapagos tortoise was soon heading for extinction. Pirates and whalers killed off huge numbers of them for fresh meat. Human occupation of the islands also brought rats, dogs and pigs which ran wild, and these wrought havoc among the eggs and the young of the armoured giants. Today only a few viable populations of the Galapagos giant tortoise remain, confined to a few, strictly protected, islands.

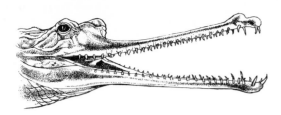

The Gavial is a fish-eating member of the crocodile order, found in the rivers of the Brahmaputra, Indus and Ganges. It will also eat any other available vertebrate and is quite distinctive, with its snout elongated into a long toothed beak with enlarged nostrils. There is only one species in the family and it can attain a length of 23ft

Crocodiles

When the zoologist and expert on crocodiles, C. A. W. Guggisberg, visited East Africa in 1952, the Nile crocodiles (*Crocodylus niloticus*) he saw in the Victoria-Nile and in the Murchison Falls National Park in Uganda were the first he had seen in the wild state. Like logs of wood that nothing can awaken, the giant armoured lizards lay in great numbers, motionless, on the river banks. In fact the beasts were extremely alert, and the least disturbance sent them crawling swiftly on their bellies into the water. Crocodiles have very sensitive faculties; they see, hear and smell extremely well. They are also marvellous swimmers, driving themselves forward with powerful strokes of their tails. But even on land crocodiles can move effectively. They stalk along, walking high on their legs, and even sometimes break into a bouncing gallop.

Crocodiles lying in the sun on a river bank often have their jaws wide open. They do this for purposes of evaporation, which helps them to keep cooler, and in this way they are able to cope with astonishing heat, up to near the critical maximum body temperature of 38°C (100°F). When the Nile crocodile closes its jaws it seems to be grinning. This is because, in crocodiles of the genus *Crocodylus*, all the upper and lower teeth remain clearly visible. In the American alligators and caymans, on the other hand, the lower teeth fit inside the upper row, and the enlarged fourth lower tooth sticks out in a most unsightly way into a pocket on the upper lip.

The size of the crocodile is often exaggerated. The longest known crocodile was killed in 1916 by Captain Riddick in Lake Kyoga, East Africa; when measured it was found to be 26ft 8in. long. The normal maximum length of the Nile crocodile is 13-15ft. The dwarf crocodile (*Osteolaumus tetraspis*), found in West Africa and the Congo area, reaches only $4\frac{1}{2}$-$5\frac{1}{2}$ft.

In the breeding season, the female Nile crocodile digs a hole in the sandy river bank one or two feet deep and lays between thirty and fifty hard-shelled eggs in it, which she then covers over again. During the incubation period of eleven to thirteen weeks the female carefully guards her 'incubator', for it is threatened by many enemies—the Nile monitor, birds of prey and predatory mammals. Just before they hatch, the baby crocodiles give clearly audible, squeaky cries, whereupon the mother crocodile uses her belly to clear away the hard-baked sand or earth covering them. At this time she is particularly aggressive and will attack any animal or man that approaches. As soon as the young have broken through their egg-shells, using the 'egg-tooth', a horny lump on the tip of the muzzle, and have scrambled out into the daylight, they make straight for the water. The mother goes with them, like a hen with her chickens, and it is absolutely necessary that she should, for the whole crocodile swarm in the river is waiting to eat up their own new generation. Crocodiles are terrible cannibals. The mother crocodile stays watching her young in the river for several days, but after that the only advice for them is: hide and get away! For after a few days the mother crocodile joins in the cannibal feast, even eating her own brood, which she has so solicitously watched over and defended for so many weeks.

Not all crocodile species lay their eggs in sand. The female of the American alligator (*Alligator mississippiensis*), of the West African dwarf or broadfronted crocodile (*Osteolaemus tetraspis*) and the South Asian estuarian crocodile (*Crocodylus porosus*), for instance, mix mud and vegetable matter in their mouths and build a kind of incubator about 3ft high and up to 6ft wide. The fermentation of the vegetable matter ensures the necessary constant warmth for the eggs to hatch.

At first young crocodiles eat insects, crabs, frogs and lizards. Later, when they have attained a length of about $7\frac{1}{2}$ft they live mainly on fish, and by the time they are $10\frac{1}{2}$ft long they take quite large mammals as they come down to the river or lakeside to drink. Crocodiles generally seize their prey by the nose, drag them under the water and kill them by drowning. The dead animal is then stored away somewhere in calm water and the crocodile waits until it rots. Its teeth are not well adapted for tearing and chewing, and it is only when the corpse is rotten that the crocodile can tear pieces off it, fastening its teeth in it and rolling over and over in a jerky way. But for most crocodiles the principal source of food is fish. The gavial (*Gavialis gangeticus*), with its long narrow muzzle, is, for example, purely a fish-eater, and so are many caymans and alligators. Stones have often been found in the stomachs of adult crocodiles; in an animal 12ft long they might weigh up to 10lbs. They presumably serve to stabilize the animal under water.

In attacking warm-blooded prey, large crocodiles seldom distinguish between animals and man. These occasional attacks on men who have approached the water without due care, or fallen into it as the result of an accident, have made the crocodile a much hated beast, especially by Europeans, and in consequence it has been ruthlessly hunted for the past 150 years and shot in vast numbers. Added to this there are now the enormous demands of the crocodile-leather industry. As a result of this relentless persecution by man, the crocodile, a Mesozoic relic surviving into the present day, suddenly faces the extinction of many of its species. An agreement just reached between representatives of nature preservation movements and of the international crocodile-leather trade in Europe, the foundation of the American Alligator Council for protection of the American alligators and caymans and the creation of vast game parks in Africa like the Lake Rudolf park in Kenya and Murchison Falls National Park in Uganda, leave us with the hope that some at least of the most seriously threatened species may still, at the eleventh hour, be preserved.

The tuatara

It caused a sensation when, in 1867, the zoologist C.L.G. Günther of the British Natural History Museum recognized a 'living fossil' in a medium-sized, olive-grey lizard from distant New Zealand. On tiny islands in a remote corner of the earth a relic had survived for 135 million years, to give

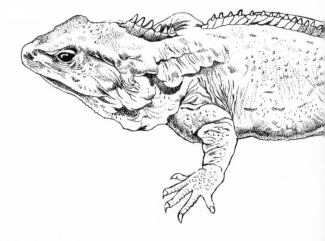

living evidence of the appearance and the behaviour of the saurians of the Triassic and Jurassic periods, 190 to 140 million years ago. A rather small piece of evidence, only about two feet long, but still a living witness, with bones, flesh, blood, nerves—and a complete set of organs; a true little saurian.

The zoologists classify tuataras in the order of Rhynchocephalia, or beak-headed reptiles which flourished in the middle era of the Earth (Triassic-Jurassic period). Today this order comprises one single family, with a single genus which consists of a single species: the tuatara (*Sphenodon punctatus*). Peculiarities in this animal's formation recall its ancestors, long since extinct: two temporal openings in the skull with a bridge of bone between them, the beak-like lengthening of the fore part of the skull, the superficial attachment of the teeth to the jawbone, the low rate of metabolism. Between the frontal and parietal bones in the fore part of the skull there is a small open gap, the site of the 'third eye' that the tuatara possesses in common with a few lizards, primitive fish and many frogs. This simply constructed eye is tiny, only 0.53mm ($\frac{1}{50}$in.) in diameter. It consists of a lens and a retina with light-sensitive cells, and is covered by the scales of the head. Scientists have been greatly puzzled by this third eye, whose object and function they have still not clarified. It may have some part to play in regulating body temperature or in the periodicity of diurnal and seasonal activity.

The tuatara occurs only on some twenty islands in the Cook Straits, off the north coast of Auckland and in the Bay of Plenty. The animals, up to 2ft long and weighing some $2\frac{1}{4}$lbs—the tail accounts for about half its total length—live in holes which they dig for themselves, or in burrows originally made by sea-birds. In the cool climate in which it lives, the tuatara, with a low rate of metabolism and a low body temperature, is ideally adapted to its circumstances. The average body temperature of active animals has been measured at 11°C (52°F), reaching a maximum of 13.3C (56°F). In all other reptiles activity begins at a minimum of 14°C (57°F). The tuatara is an animal of the twilight and the night. Its beautiful, big eyes, with a perpendicular slit pupil, enable it to hunt for insects, earthworms, snails, birds' eggs, young birds and small lizards in the half-dark.

The tuatara has another peculiarity in common with the ancient creatures: it has no sexual organ. To pass semen into the female in the act of mating the two animals press their cloacae firmly together. In the southern spring (October-December) the female lays eight to fourteen parchment-shelled, blunt-shaped eggs in a hole which she digs for herself. The incubating period lasts for from thirteen to fifteen months, so that the hatchlings have to withstand a hibernation inside the egg. After leaving the egg the young lizards grow only very slowly; a tuatara does not become sexually mature for something like twenty years. Growing as slowly as they do, they enjoy a long life. Tuataras have been kept in captivity for seventy-seven years, but it is reckoned that the actual life expectancy is 120-150 years.

The name 'tuatara' is a Maori word, meaning 'peaks on the back'. It refers

The tuatara is isolated in time and place. Only found today in New Zealand, its closest relatives died out 100 million years ago. It grows abnormally slowly but may live to 50 years and reach a length of 2ft. Its main food is insects, though it will occasionally take a young bird

to the spinal comb, specially marked on the male, which can be erected when the animal is provoked and hangs loosely to the side when it is at rest. There were tuataras on the New Zealand mainland until the middle of the eighteenth century, but the introduction of foreign mammals such as cats, dogs, pigs and weasels led to their extinction. Only if their home on their tiny islands can be kept free from these invaders will it be possible to preserve these ancient and interesting reptiles in times to come.

Chameleons

A nature-lover travelling in tropical or southern Africa may, with good luck and a bit of patience, get a chance to watch a flap-necked chameleon (*Chamaeleo dilepis*) in the branches of a bush. The animal's movements look sluggish as it clambers along a branch, its feet slowly letting go and reaching forward for a new foothold. The two outer and three inner toes of a chameleon's fore feet, and the three outer and two inner toes of its hind feet, grow bunched together so that they form pincers, and it uses its long tail, often rolled up, as an extra limb for holding on. As it creeps along, the chameleon looks the very embodiment of sleepiness and apathy. But the appearance is deceptive. You can tell from its eyes, constantly turning from side to side, that the creature is paying lively and alert attention to what goes on around it. The eyes of a chameleon are large and spherical, and each one can be turned independently of the other. The pupil forms a little hole in the centre of the scaly eyeball. An insect attracts the attention of our flap-necked chameleon in the bush. The eyeballs turn forward, staring at the coveted prey. While previously the chameleon was taking in its surroundings monocularly, each eye continually wandering, now it focuses binocularly on the insect, ready to shoot out its tongue and catch it. Its mouth is half open, the tip of the tongue gradually appearing between the lips—and then suddenly the tongue, as long as the animal's body, with a broad, club-shaped tip, shoots out with a snap like a bowstring. The thick, lobed tip, made sticky by a secretion of the salivary gland, is wrapped round the prey like an elephant's trunk and whisks it back into the mouth. This action of the chameleon's tongue is quick as lightning; it takes no more than $\frac{1}{25}$ of a second. Chameleons are unsociable and jealous of their solitude. If another chameleon appears among the branches it is quickly dealt with. Our chameleon spots its rival in an instant and becomes greatly agitated, spreading out the skinny flaps at the back of its head and beginning to change the colour of its body. The basic colour, light green or bark-coloured, now turns darkest green and is covered with yellowish and white spots. Chameleons are absolute masters of colour change; they can express every passing emotion by a change in their colours and markings. In addition to their light-sensitive skin they mainly rely on their eyes as the means of control.

Hoehnel's chameleon too (*Chamaeleo hoehnelii*), in East Africa, changes colour in extreme agitation at the sight of a rival. Its usual brown or brownish-

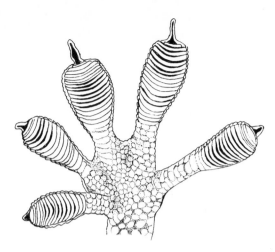

The tokee (Gecko gecko) of South and South-East Asia, like the majority of gecko species, has adhesive pads on its toes, furnished with microscopic hook-cells, which enable it to run up walls and along ceilings with no difficulty. All geckos have calls, in the case of the tokee a loud baying which can startle an unsuspecting passer-by

black turns to a whole range of green colours. When two males meet they circle one another threateningly, until finally one of them goes dark and so acknowledges defeat.

Also in East Africa, Jackson's chameleon (*Chamaeleo jacksonii*) has three remarkably large horns on its snout. The mountain chameleon (*Chamaeleo montium*), found in Cameroon and on the island of Fernando Po, has two long horns on the tip of its snout and a seam of skin like a fin along its back as far as the root of the tail. It displays a particularly fine play of colour when disturbed. The shining leaf-green suddenly goes light yellowish-green and the front of its body, especially its head, is covered with a network of whitish and blue spots. If a human observer, for example, approaches, the mountain chameleon plays a remarkable game of hide and seek. It moves to the side of the branch away from the observer and there, camouflaged by its slender shape and green colour, it completely vanishes.

Monitors

If you come across the giant Komodo dragon (*Varanus komodoensis*), either in the Zoo or in its homeland in the island of Komodo, you think you are looking at a real dragon out of a fairy tale. Ten feet long, huge and light-coloured, the animal can weigh as much as 300lbs, yet it moves with astonishing grace. This largest of all lizards after the crocodiles was first discovered in 1912 by Major P. A. Ouwens, director of the Botanical Gardens in Buitenzorg in Java. Everything about the animal is gigantic, even its eggs, which can be up to 5in. long and weigh almost 8oz. Although it is strictly protected in its homeland, the number of Komodo dragons is falling, mainly as a result of the persistent hunting of its principal food, the pig-deer and the wild pig.

Herodotus, writing in the fourth century AD, took monitors to be crocodiles; he described the desert monitor (*Varanus griseus*) as a 'land crocodile'. Monitors, we may say, are the most intelligent of lizards; they can even be trained to some extent. They have a well-developed sense of direction, a remarkable ability to find their way, and it has been known for a monitor to recognize its keeper and answer to its name. Most monitors are good climbers, and many swim and dive marvellously. The Nile monitor (*Varanus niloticus*), for example, can stay under water for up to an hour; it swims, arms and legs pressed against the body, with vigorous winding movements of its rump and tail. Monitors have strong jaws and sharp teeth; for defence they use the razor-sharp claws on both fore and hind feet. They are carnivorous and feed on anything from insects, birds and birds' eggs up to quite large vertebrates. They shake their prey to death like a terrier, or beat it on the ground with violent movements of their head.

The monitor family can be found all over the tropical and sub-tropical areas of Africa, the Near East and southern Asia, in the Indo-Asiatic islands and in Australia. Particularly lovely is the emerald monitor (*Varanus prasinus*) from New Guinea, with its bright green colouring. It lives mostly in trees, and has a true prehensile tail. The two-banded monitor (*Varanus salvator*) of southern Asia is a magnificent dragon, living in damp forests or on wooded river banks, never far from water. When two males of the Bengal monitor (*Varanus bengalensis*) from southern Asia fight, they do it according to a formal set of rules, standing up on their hind legs and seizing one another with their forelegs in an effort to throw their opponent on to his back. These fights between two rival males are called 'ritual battles'.

The greatest number of monitor species is found in Australia. Among them is the lace monitor (*Varanus varius*), with its irregular pattern of narrow, light spots and stripes; the mangrove monitor (*Varanus semiremex*), which lives in trees, generally mangroves, in Queensland and northern Australia; the little desert pygmy monitor (*Varanus eremius*), which lives in holes in the ground; and the short-tailed pygmy monitor (*Varanus brevicauda*), which, at about 8in. from head to tail, is the smallest in this family of great dragons.

Gaboon viper and tropical rattlesnake

In the primeval forests of tropical Africa, from Cameroon over the whole Congo basin as far as Uganda, western Kenya and south-western Tanzania, you will find the Gaboon viper. In the terrarium this squat snake, thicker than a man's arm, looks too brightly coloured to miss, like an oriental carpet. But on the ground in the forest, when the sun's rays fall through the roof of leaves to paint a jumbled pattern of light and dark shapes, the bright pattern disappears completely. The colours seem to dissolve; the snake merges into its background.

The Gaboon viper grows up to 6ft long; females are generally larger than males. The snake has a quiet, almost phlegmatic nature and bites only seldom, if disturbed or excited. When it is hunting, however, it can strike so quickly that the movement can hardly be seen. It sinks its two huge poison-fangs, from ¾in. to 1½in. long, deep into its prey and, as with a syringe, injects a great quantity of extremely powerful poison into the wound. The poison contains both haematoxic elements, which destroy the blood and vascular system, and a high proportion of nerve-poisons, the so-called neurotoxins. This high proportion of nerve-poisons is unusual in most of the vipers. The poison-fangs are set on a hinged bone, so that they can be protruded when striking; immediately after biting, the Gaboon viper withdraws its fangs and waits for the poison to take its deadly effect. Small animals killed for food, however, it will hold on to, and finally swallow them head first. In doing so it uses its poison fangs, which it can move right and left, separately and independently, to push the mouse, rat, bird or toad down from behind with an action like an excavator, until it has disappeared. The meal over, the poison-fangs are carefully returned to their mucus-lined pouches, bedded down and stowed away, with elaborately graceful movements of the jaws.

Gaboon vipers bear their young alive (not as eggs), the female producing

The open mouth of the common viper or adder clearly shows the poison fangs. Here they are paired, as sometimes happens. In the viper family (which includes rattlesnakes) the fangs are long and lie flat against the palate when the jaws are closed, only becoming erect when the jaws open. As the snake bites, poison is injected from the poison glands through the hollow fangs into the wound. The forked tongue is a sensory organ for smelling and touching. Though not generally fatal, the bite of a viper is extremely painful and a few deaths are reported every year

from thirty to fifty young. This snake's method of locomotion is very unusual. It presses its ribs, which are flexibly attached to the spine, down on to the scales of its belly and pushes itself forward, 'walking on its ribs'. It looks like a children's sack-race. Its relatives, the puff-adder (*Bitis aritans*), and the even more brightly and beautifully coloured rhinoceros-horned viper (*Bitis nasicornis*), can move in the same way. The little horned puff-adder (*Bitis caudalis*), from the desert areas of southern Africa, is in contrast a side-winder, darting sideways across the sandy soil and able to bury itself in the sand with shovelling movements of its ribs.

As the Gaboon viper is the most fearsome of the African snakes, and has the most powerful poison, so the tropical rattlesnake or cascaval (*Crotalus durissus*), found in Central and South America, is of all American rattlesnakes specially feared for its deadly bite. Its poison, too, contains a high proportion of nerve-poison compared with the typical rattlesnakes' blood-poison. The tropical rattlesnake belongs to the family of pit-vipers (Crotalidae), the New World equivalent of the Old World vipers. Snakes of this family have a groove between eyes and nostrils, the site of a membrane which is highly sensitive to temperature. With its aid the snakes can locate warm-blooded creatures even in total darkness, creeping up on the scarcely perceptible source of heat. Rattlesnakes get their name from the chain of overlapping, hard, horny rings at the end of the tail, which are the remains of previously sloughed skins. A new ring is formed every time the snake sheds its skin, but the end rings break off from time to time so that the number of rings stays between six and ten. When the snake raises its tail high in the air and vibrates it from side to side the rings make a rattling noise that can be heard over a distance of more than 50 yards. The rattle is an effective warning to any potential enemy; it corresponds to the pumping hiss given by the Gaboon viper when roused.

Snakes that climb trees

Snakes can not only swim for long periods, glide over the desert sand forwards and sideways or crawl swiftly through the grassland, they are also very skilful climbers. The climbing snakes of Europe are mostly members of the harmless rat and chicken snakes (genus *Elaphe*), such as the Aesculapian snake (*Elaphe longissima*) and the pretty coloured common leopard snake (*Elaphe situla*), which readily climb up walls, bushes or trees. Their bodies are thin, and slightly keeled ventral scales give them an angular shape along both sides of the underneath of the belly, which helps them to get a better grip on rough, sloping surfaces. These snakes kill their prey by winding themselves round it and throttling it before swallowing it.

Angled ventral scales running along both sides of the under-belly testify that the spotted rat snake (*Spilotes pullatus*), found in Central and South America, is another real tree climber. It eats birds, mice, lizards and even other snakes.

There is a specially large number of climbing tree-snakes among the rear-fanged snakes of the sub-family *Boiginae*. The members of this sub-family have poison-glands, and the back teeth in their upper jaws are extended and have a groove along their front side. The poison of the rear-fanged snakes is generally weak, just strong enough to kill the small animals on which they feed. The great majority of snakes in this group are thus harmless to mankind; only the bites of the grey twig-snake (*Thelotornis kirtlandii*) and the boom-slang (*Disopholidus typus*) both found in Africa, have ever proved fatal to humans. The Asian green whip snake (*Ahaetulla mycterizans*) of South-East Asia and Indonesia is one of the rear-fanged snakes that climbs trees. This snake is very thin and its upper side is leaf-green. This long, thin, outward appearance, good camouflage among the branches, is emphasized by horizontal oblong pupils, a light stripe running back from the nostril over the eyes and a mobile tip to the snout like a turned-up trunk. The green whip snake is very light, so that it can lie out on the thinnest twigs; the front part of its body hangs down, swaying slowly to and fro, and it fixes its eyes firmly on its prey before it strikes.

There are other tree-snakes with horizontal pupils, among them the American green whip snake (*Oxybelis fulgidus*) of Central America, whose back is light green with two pale stripes, and the grey twig-snake, found in Central Africa. This grey tree-snake inflates its neck like a pouch when threatened, making its head look enormously bigger. When it hangs in the branches its head and the front of its body hang down and swing jerkily to and fro. At the same time it puts out its tongue, bright red with two points of black at the double tip, and this acts as a bait. Tree-frogs and geckos are attracted by it, then suddenly seized and held fast until the poison runs down the grooved fangs at the back of the mouth into the bite and takes effect.

The ornament tree snakes of the genus *Chrysopelea* live in South-east Asia and Indonesia. One of them, the paradise tree-snake. (*Chrysopelea paradisi*), is sometimes called the 'flying snake', although it cannot actually fly or glide. Certainly, when striking or diving from a tall tree, it does flatten its body, spreading its ribs wide so that the ventral bones are arched concavely inwards to form a furrow along its body, which makes it look as if the snake was drawing in its stomach. This does presumably enable the snake to slow down its dive a little, so we can properly speak of a 'parachutist effect'. It is also possible that the spread ribs work like springs to soften the snake's landing from a dive. The paradise tree-snake likes to climb up coconut-palms, where it hunts for geckos and tree-skinks. When one of these snakes in the top of a coconut-palm wants to move on, perhaps to go and hunt somewhere else, the easiest way for it to leave the tree top would obviously be by a jump into space controlled in this way.

Right The Komodo dragon is the largest of the monitors, a genus of large lizards. It may grow to 10ft. Only found in the island of Komodo and a few other islands in the Sunda group, Indonesia, it is a flesheater, capable of killing even goats and small deer

Overleaf True alligators are only found in North America and China and are cousins of the crocodile. Found in the Mississippi, they have been known to grow to 19ft. Alligators, their close relatives caymans, and crocodiles all have enlarged fourth teeth in their lower jaws, but in the alligators and caymans this slots into a socket in the upper jaw and is invisible when the jaws are closed. The fourth tooth of the crocodile is clearly visible even when the jaws are closed. Alligators feed on mammals and fishes but their numbers are dwindling since their valuable skins are used for purses and other articles and their teeth for ivory

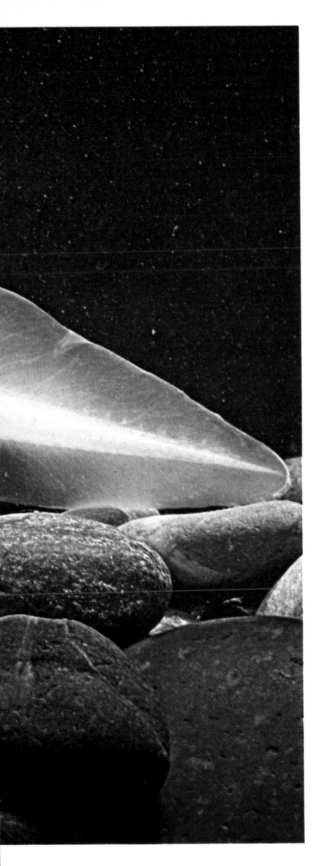

Left The strange-looking Mexican axolotl is a type of salamander, but it never normally develops beyond the larval stage. However, even in this non-adult condition it can lay fertile eggs. This specimen is a not uncommon albino variety, with its pink, larval, external gills feathering out behind its head. If the water in which they normally live slowly dries up, axolotls lose their gills and take on the form of typical salamanders

Above Chameleons, like this flap-eared species from South Africa, are slow movers, stalking their prey under the protection of their marvellous camouflage. When the prey is within striking distance, the long, sticky tongue, with its club-like tip, flashes out and back again within $\frac{1}{25}$sec., and the unsuspecting insect is swallowed

After copulating at sea (*above*), huge numbers of Pacific Ridley turtles come ashore on the coasts of Central America (*right*) to lay their eggs in pits in the sand (*top left, opposite page*). The young turtles all hatch at the same time (*top right, opposite page*) and make a floundering dash for the sea (*opposite bottom*). Thousands never get there, for they are easy prey for waiting birds

Fishes

by Alwyne C. Wheeler

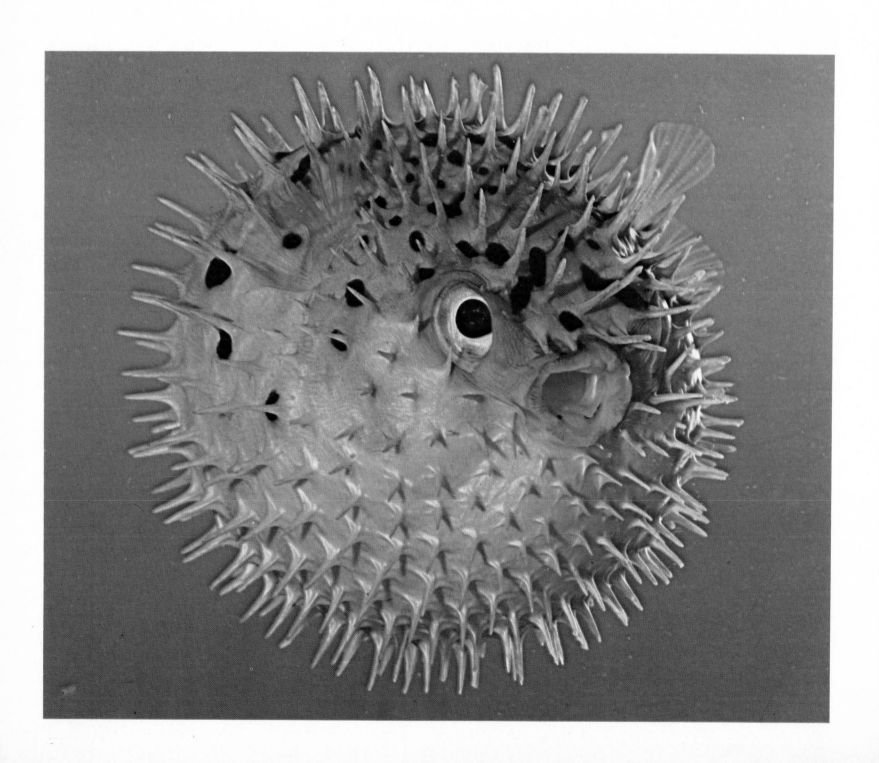

Introduction to fishes

Fishes can be broadly defined as aquatic, cold-blooded, vertebrate animals, which breathe by means of gills and propel themselves by fins supported on bony skeletons. As with all definitions, the exceptions spring immediately to mind: a few fish can survive long periods out of water, some are distinctly warm-blooded, others have supplementary lung-like organs and breathe air very efficiently, while in others the fins play little part in locomotion. The exceptions, however, go some way to indicating the great wealth of variation within the world of fishes, and the definition stands for the great majority of fishes.

There are probably around 22,000 different kinds of fishes living today. The great majority of them are bony fishes of the class Teleostomi, but there are about 600 species of cartilaginous fishes (sharks, rays and chimaeras), and rather less than fifty lampreys and hagfishes, the most primitive of the fish-like vertebrates. The teleosts or bony fishes have successfully exploited most of the aquatic living spaces of the world. They are found in tumbling hill-streams, in mountain lochs, and in the deepest trenches of the oceans, in equatorial swamps, and under the polar ice. Virtually no sizable water is naturally uninhabited by fishes, the only exceptions being excessively salt lakes or seas (such as the Dead Sea) and the depths of the Black Sea, where due to local conditions there is little or no dissolved oxygen to support life of any kind.

Not surprisingly, in view of the immensity of the living space available, the bony fishes show a tremendous variety of adaptations. Many species living on sea shores have developed suction pads by which they cling to rocks in the swirling surf; small examples are the cling-fishes and sea-snails, while others such as the lumpsucker grow to a length of 24in. Hill-stream fishes face the same problem of being swept away by the force of water and, although in no way related to these, marine fishes have evolved similar suction devices. The Asiatic hill-stream homalopterid loaches, for example, have a broad flattened underside and expanded pectoral and pelvic fins, which together turn the ventral surface into a powerful sucker. Other fishes living in running water or in wave-swept situations have long, compressed bodies, and seek safety by burrowing between rocks or into the river bed.

The open sea has the greatest wealth of fishes. Most of the smaller fishes are shoaling species, such as herrings, sardines and anchovies, which ensure their survival by forming shoals, feeding on minute planktonic organisms,

Puffer fish, like this porcupine fish, have spines instead of scales. When defending themselves they take in air or water and inflate to an impenetrable spiny ball. They are slow swimmers, relying on their appearance and spines for defence

Sharks, rays, skates and chimaeras belong to the class of cartilaginous fishes. This shark is quite representative of the class. The skeleton is basically formed of gristly cartilage. The gills have unprotected external openings and the teeth are in a band attached to each jaw. The scales are small and tooth-like. Not particularly fast as a swimmer, the shark is more likely to prey on sick or sluggish fish

and being basically silvery in colour so that they merge into the ceiling of their world, the mirror surface of the sea when lit by the sun. Even some of the giant fishes of the ocean feed on plankton, including the basking shark, the whale shark, and the manta rays, all well-equipped to filter from the water the hordes of small crustaceans, young fishes and larval molluscs and crustaceans which form the plankton. Other large surface-living fishes, such as the tunas and billfishes, feed extensively on the anchovies and sardines, and are the great swimmers of the sea, fast enough to attack the protective shoals. They are also migratory, at times crossing the ocean basins.

Below the well-lit surface waters, lantern-fishes and bristle-mouths replace the shoaling anchovies. The former are liberally equipped with distinctive light organs, enabling the shoals to keep together and species to recognize one another. Lantern-fishes, and other families, make extensive daily vertical migrations (as do many other mid-water organisms), rising near to the surface at night and sinking through the twilight zone as the sun rises. In the depths there are a variety of predators preying on these fish, many sharks, sabre-toothed dragon fishes, star-eaters, deep-water angler fishes and others, often armed with massive fangs and capacious jaws and some of them equipped with light-organs acting as luminous lures. The sea-bed in the depths of the ocean is sparsely populated by fishes, but tripod fish have been observed, perched on their long fin-rays, perhaps in this way keeping clear of the bottom mud, and halosaurs and deep-water cods snake along just above the bottom in a characteristic head-tilted-down posture searching for edible items. Several kinds of brotulid have also been found.

Life in the ocean depths must be rigorous but numerous fishes have adapted to it. Similarly, others have adapted to rigorous conditions in fresh-water. African swamps are inhabited by several species of fish which live in the deoxygenated water of the warm season by breathing air, most notable of all being the several kinds of lungfish which survive months of drought in a dry burrow. Several species of rivulin, beautifully coloured and often only two inches in length, live in ephemeral pools, laying eggs which survive drying out in the dry season but which hatch to form the next generation.

Most fishes fall into the class of bony fish. Salmon, eels, herrings, flying fish, carp, even the coelacanth, all have bony skeletons. Many have a swim bladder, which helps to keep them afloat. There are two sub-classes: the fleshy finned fishes, which include lungfishes and the coelacanth and which were the forerunners of amphibians; and the ray-finned fishes, which contain most of the food fish, including herrings, salmon and cod

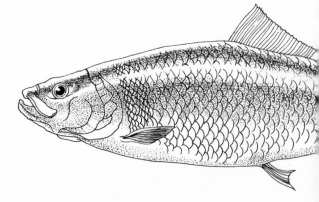

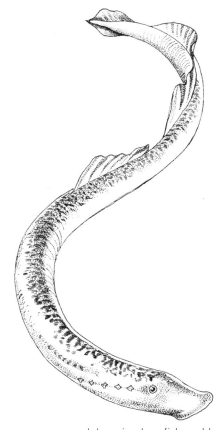

Lampreys are scaleless, jawless fish and belong to the Class Agnatha, the oldest Class of vertebrates. They are parasitic on other fish, to whom they cling by their round, tooth-studded mouth, rasping at the flesh with their tongue and draining their hosts of blood. They also scavenge. Adults die after spawing, which takes place after a migration up rivers and streams. The larvae are not at first equipped for a parasitic existence, and feed off bottom-living micro-organisms, but they eventually metamorphose into adults

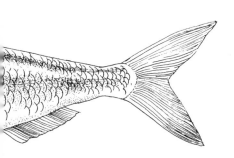

Related toothcarps live in arid conditions in North America, such as the Death Valley system of eastern California and south-western Nevada, where a small pupfish lives in summer time in isolated shallow pools with a salt content twice as high as normal sea water.

The ability to live in such habitats demonstrates the capacity of fishes to adapt to all kinds of living conditions and partly explains their tremendous success as animals. Despite their ancient lineage, for they are the oldest and most primitively organized of all vertebrate groups, they are more abundant in species and, within the limits of their watery world, more widely distributed than any other vertebrate group living today. Their success also accounts for their fascination, for simply on the grounds of numbers and the inaccessibility of their habitats to exploration, they are the least known of all the higher animals.

Strange associations

The habits of the anemonefishes, of which several species are known in the tropical Indo-Pacific, are well known today through observations by divers and study in the aquarium. These fishes are mostly small (up to 4 inches in length) and boldly coloured; they can be found living in association with large sea anemones on coral reefs. Several of the known species will inhabit only one species of anemone although others are not so choosy. The fishes when adult are usually found in pairs, living in an area of reef which includes the sea anemone, in which they shelter if threatened. Normally contact with the stinging cells in the anemone's tentacles means instant paralysis and death for a fish within minutes, but the anemonefishes dart amongst the tentacles with no ill-effect. Their skins are believed to secrete a mucus or slime which inhibits the action of the stinging cells.

The precise relationship between the anemone and the fish is not certainly known. It is thought that the fishes derive some measure of protection both when they hide within the tentacle mass and when they lay their eggs on the coral beside the anemone. Do they pay for their keep? It is sometimes claimed that the presence of the fish may lure smaller fishes towards the anemone and their death but, although this is possible, it is not supported by observations.

However, it is likely that they protect the anemone from the coral polyp-browsing butterflyfishes, and the coral-grazing parrotfishes of the reef, for the anemonefish will ferociously attack any intruder within their territory.

Numerous other associations between fishes and coelenterates are known. The small stromateid fish, *Nomeus gronovii*, lives, at least when young, amongst the trailing tentacles of the siphonophore *Physalia*—the Portuguese man-of-war. The stinging cells in these tantacles can raise a weal like a whip-lash on a man's hand, yet *Nomeus* lives amongst them unharmed! The fish enjoys some protection from this association, but it is said to browse on the tentacles. On the other hand some Portuguese men-of-war have been found with partly digested *Nomeus* in their grasp, so the association is clearly complex and not fully understood yet.

Several members of the cod family similarly associate with jelly-fishes. The young cod, haddock and whiting in the North Atlantic, and walleye pollack in the North Pacific have all been found swimming close beneath the bell of medusae of several different kinds. Again the advantages of doing so are not certainly established but the young fishes probably gain some protection, and are hidden from aerial predators such as terns.

Hat-pin sea urchins (*Diadema*), with their long, black spines, are familiar objects in tropical and warm seas, and with the exception of certain trigger-fishes they are unmolested by most marine animals. Several kinds of fish have taken advantage of their immunity and enjoy a symbiotic relationship with the sea urchin. The best known is the little shrimp-fish, which is widely distributed in the Indo-Pacific, a slender, very compressed fish with a long pointed snout. The shrimp-fish habitually swims in a head-down posture, and has narrow black lines running from the tip of the snout to the tail along each side and on the back. This colouring makes the fish well-nigh invisible against the long black spines and the protection it derives must be considerable.

Fish migrations

Migration is part of the life style of many kinds of fish, but with the exception of some of the commercially exploited food fishes we know very little of the subject. Essentially, migrations are associated with the availability of food and with the reproductive cycle of the species, the former being superimposed on the latter. Many migrations are seasonal, such as the immigration into British waters of the blue shark, the mako shark and the trigger-fish, which all occur in the western English Channel and along the western seaboard in summer and autumn, attracted, in the case of the sharks, possibly by the shoals of coastal prey fish. These, and other species, are found far to the south-west in the warmer Atlantic during the winter months.

A comparable migration is made by the Atlantic blue-fin tuna or tunny, which spawns in the Mediterranean and off south-western Spain. Juvenile tuna are encountered in the northern North Sea from June to October when they disappear, presumably having swum up the west coast of the British Isles in spring to feed on the sprats, sand-eels, capelin, and other fish so abundant in the north. Large tuna formerly occurred in the North Sea and followed the shoals of herring on their migrations, but these, like the younger fish more recently, were ruthlessly exploited and are now greatly reduced in numbers. The finding of fishing hooks of Turkish and eastern Mediterranean manufacture in occasional large tuna showed that they did indeed make the journey from the Mediterranean to the North Sea. More recently, a number of large fish have been captured bearing numbered tags fastened by American fisheries workers in the western Atlantic. This suggests that the migration of the European blue-fin tuna is supplemented by trans-Atlantic migration,

The migration of the Atlantic salmon
The complex migrations of the Atlantic salmon are still not fully understood, but it is known that those spawning in British rivers have made migrations from the Atlantic south of Greenland and from the Arctic waters of Norway. After the young fish have hatched in the streams and rivers, they will eventually make their way back over the migration routes. Similarly salmon from the rivers of eastern Canada and the United States also make the migration to and from the waters south of Greenland

and that the situation is more complex than at one time was realized.

The oceanic migrations of the Atlantic salmon are likewise now known to be similarly complex. The salmon spawns in the upper reaches of rivers on gravel beds, and the young fish after two or more years in the river descends to the sea. At one time it was thought that they stayed in coastal waters feeding on marine organisms to return to the river of their birth after the passage of one or several years. However, it is now known that they make extensive migrations involving (in the case of many European salmon) an Atlantic crossing to the waters of southern Greenland. This was first demonstrated by the development of a fishery in the open sea off the Greenland coast which used light floating nets at the surface. The huge catches there were clearly more than the rivers of north-eastern Canada and Greenland could support, and the capture of fish marked in Scotland and Ireland with fish tags showed that some had crossed the Atlantic. A similar fishery later became established in the Norwegian Sea, and it is now accepted that the salmon which spawn in a British river will have journeyed from the stream of their birth far out across the Atlantic or up to the Arctic waters of Norway before returning to spawn.

The migration of salmon is thus far more complex than that of many fishes, and as its details are unravelled, the mechanisms involved become increasingly remarkable. It is obvious that migration is not simply a random movement, and the Atlantic crossing can only be achieved using an efficient navigation system. Experiments with salmon have shown that they have a sun-compass sense of direction, and it is likely that they are able to navigate by night as well as by day. Sun-compass navigation implies not just a general awareness of the position of the sun, but the ability to make corrections for its changing position relative to the earth as the day passes, and also for its seasonal position. Navigation by this means is sufficient to help the returning salmon to the general area from which it originated, for example from East Greenland to the coast of Britain, but not to bring it to its natal stream. This is effected by an acute sense of smell, which is capable of detecting the individual odour of the river and even stream of its birth, an odour which must have been imprinted on the memory of the fish very soon after hatching.

A four-eyed fish and other fishes' eyes

In the coastal waters of Central and north-eastern South America there lives a fish which has four eyes. Living in shallow muddy coastal waters, estuaries, lagoons and lakes, it lies at the very surface with the upper parts of its head out of the water. Each eye is protuberant and is divided across the centre in the line of the body axis by a band of opaque tissue which effectively divides the eye into two separate eyes. Not only is each eye physically divided but each half-eye has a separate retina, and the lens of the upper part is more rounded than that of the lower. As the fish maintains its position at the surface the upper half-eye is exposed to the air, and with its rounded lens is well equipped to detect aerial prey or predators. Simultaneously the lower half-eye can scan the aquatic environment.

While this represents an extremity of modification amongst fishes' eyes, even 'ordinary' fish eyes are beautifully adapted. The eyes of all vertebrates have broad similarities, but there are differences between the eyes of fishes and those of mammals. First, the lens in a fish's eye is a perfect sphere, and it is placed close to the bulging surface of the eye. The lens therefore lies partly outside the profile of the head and provides the nearly all-round vision that a fish requires in the absence of a head movable on a neck. In addition, studies of fish eyes have shown that they are long sighted. In the lateral field their long-sightedness is most pronounced, while forwards their vision is normal to even short sighted. The advantages of this complex situation are essentially that forwards a fish needs to see its prey or a mate at close quarters, literally at the end of its snout, but laterally it is more useful to detect movements at a distance, possibly a predator, than it is to see it in detail close up.

Eyes also vary with the habits of the fish. Nocturnal fishes have big eyes, the retinas of which have very large numbers of cells known as rods which are sensitive to extremely low light intensity. Diurnal fishes have fewer rods but more cones in the retina; the cones are less sensitive but register the bright colours and well-lit shapes. They also have relatively small eyes.

Cave-dwelling blind fishes are known from all the continents and in many families, although the greatest number are catfishes. Some, such as the Mexican cave-fish (*Anoptichthys jordani*) are quite common and are frequently kept as aquarium specimens. Like most cave-fish it has no pigment in its skin, and is a pale uniform pink with a silvery sheen. At hatching, the fry have distinct and apparently normal eyes, but these later become distorted, fail to grow and are eventually embedded in the head tissues. This species is found in three caves of San Luis Potosi, Mexico, in streams and pools. Apart from its colouration and its feebly developed eyes, it closely resembles the Mexican tetra (*Astyanax mexicanus*), which lives in the same region but possesses normal eyes and pigment.

The unsilent sea

In the early days of underwater exploration it was customary to refer to the sea as silent, but the falseness of this title was demonstrated by the development of underwater listening devices during and after World War II. These sensitive devices showed that far from being silent, the sea was filled with a cacaphony of sound produced by a variety of animals. Fishes contribute a great deal to the sonic effects of the sea, and they produce sounds by several methods.

Active, shoaling fishes make noises when they are swimming swiftly. Swimming idly or stationary in an aquarium anchovies made no noise, but directly they began to swim actively sounds were produced. The advantage in this is obvious for it means a shoal can remain cohesive even at night, and even a blinded anchovy could keep its place in experimental conditions. The noises are believed to be involuntary and are produced essentially by the swimming action.

Better known sonic fishes are the members of the mainly marine fish family Sciaenidae. Many of them are known as croakers or drums for the noises they produce. They have a large and very elaborate swim-bladder which acts as a resonating chamber amplifying the noises made by adjacent drumming muscles. The croakers evidently make full use of their sonic ability for as night falls the level of noise increases markedly. Also when imprisoned in an aquarium, they drum continuously until their surroundings have been fully explored and afterwards rarely make much noise. Many sciaenid species live in estuaries and in inshore waters where visibility is poor, and the production of sounds to keep in touch with their fellows as well as to navigate is clearly advantageous.

A similar function can be ascribed to the swimbladders of the deep-water fishes, the macrurids or rat-tails. Again the swimbladder is large, and well-developed drumming muscles connect to it in male fishes. Presumably the females are condemned to a life of silence, although clearly they can locate the males in the darkness of the deep sea by responding to their calls.

Toad-fishes, which are well known on the Atlantic coast of North America, are capable of uttering very loud calls. Again the swimbladder is large and equipped with special muscles, but their performance is formidable for a fish of 15 in. One scientist has measured the output of a toad-fish at 24in. as reaching an intensity of 100 decibels, comparable to the noise of a subway train! Toad-fishes are strongly territorial and it is thought that these fog-horn bellows are probably reserved for an encroaching rival, although they are also capable of numerous grunts, groans and boops of more modest volume.

It seems that fishes, in general, use their wide capacity for making noises much as do other animals. Low-level noises inform their neighbours of their presence and help to keep in touch, with occasional outbursts, as emotions of fear, anger or love require expression.

Colours and patterns

Brilliance of colouring amongst fishes probably reaches a peak in the coral reef communities of tropical seas. Here a whole range of often unrelated fishes display dazzling primary colours and bold patterns in bewildering variety. Most distinctive are the butterfly-fishes, deep-bodied fishes which feed mainly on the coral polyps and minute animals concealed in the creviced coral. Amongst them distinctive patterns are commonplace; bold dark bars run across the body, disrupting the outline of the fish. Many have at least a single bold stripe across the head completely concealing the eye, and others, like the four-eye butterflyfish of the Caribbean, have a conspicuous white-ringed eyespot near the tail fin. Such eyespots are common amongst butterfly-fishes in general. Clearly these fish are well protected by their markings, despite their brilliant colouration. The eye, which is always a conspicuous organ in any animal, is concealed, and a large false eye provided to attract attention. Any attacking predator would expect its prey to move forward, eyespot end first; instead, the butterflyfish will dart off in the other direction.

The brilliant colouration of these and other reef fishes always attracts comment. At first sight they appear to offer little if any protection, but viewed underwater, against the colourful living coral, the fish are by no means so obvious as they might seem when out of water.

Most fishes, however, seek to conceal themselves by subdued or cryptic colouration. Thus, flatfishes such as the turbot and dab, can vary the tone of their coloured side quickly and effectively, to merge with the background colouring. A change of position from light to dark background will be matched within minutes by the fish. Anglerfishes and frogfishes are equally expert at blending into the background, and both are well provided with small fleshy flaps of skin which help to break up the outline of the fish, and also sway gently with movement of the water, thus adding to the illusion that they are part of the background. The cryptic life-style of both anglerfishes and frogfishes is not, in the first instance, a defence against predators, it is rather a matter of concealment from approaching prey, for they all have a fishing lure with which they entice smaller fishes to within snapping distance.

Even those fishes coloured like the herring or bleak, green or blue on the back, brilliantly silvery ventrally, are well concealed. They live in the well-lit surface waters, where predators from below find them hard to distinguish against the silvery surface of the sea, while airborne predators are baffled by the dark back matching the dark coloured deeper water. This countershading is a highly effective concealment for fishes which must live in the glare of the sun. Just how well it works can be judged by the tropical African catfishes of the family Mochocidae which habitually swim on their backs, some of them feeding on the algae growing on the undersides of floating lily-pads. These upside-down catfishes have dark bellies and light backs.

Bright colours and bold patterns are also used as warning signs. The dragon fish or turkey fish of the Indo-Pacific has narrow red, black and white vertical bars over its body. This is one of the few fishes which does not flee when approached by a diver. Its back is armed with long, envenomed spines, and it is clearly quite fearless and advertises its ability to defend itself by its bold colouring and behaviour. The common Mediterranean electric ray, which can give a powerful electric shock when touched, has a striking cluster of blue, black, and yellow roundels on its back. Most electric rays are dully coloured and bury themselves in the sea bed, but this species lives in shallow water and is frequently visible out in the open. It seems likely that this bold colour pattern is a warning to potential predators. Although no connection has been established, several harmless rays in the Mediterranean have similar roundels on their backs, and it is possible that they are mimicking their dangerous relative.

Spears, stings and shocks

In general, most fishes are retiring and inoffensive in their life-style, preferring to hide or not be seen or, if visually very obvious, then being so for reasons such as the need to be seen. However, there are some fishes which are equipped with powerful weapons of defence; offensively so when they come into contact with man.

Stingrays (family Dasyatidae) are large, flattened rays which get their name from the long serrated spear at the base of the tail. They are world-wide in distribution and although most common in warm tropical seas occur in temperate waters seasonally; the stingray (*Dasyatis pastinaca*) is common in the English Channel and southern North Sea each summer. Most stingrays feed on shellfish, mussels, oysters, crabs and shrimps, and some are serious pests. They are greatly feared by fishermen, for if trodden upon their immediate reaction is to curl their tail up over their back and then with a sudden lunge, jab the sting into the man's leg. The sting, even in the relatively small British species, can be six inches long, and is equipped with venomous tissue on the lower side. The wound is excruciatingly painful, becomes inflamed, often infected, and is always serious. In Australian seas, stingrays attain a width across the disc of six feet, and wounds from these huge fish have caused several deaths.

More modest in size, but in many ways more dangerous, the stonefishes (*Synanceia* spp.) are also found in Australian tropical waters, and across the Indian and most of the Pacific Oceans. Their appearance is grotesque, squat with a broad head; and the skin is covered with warts and dull blotched brown in colour. Both the common species live in shallow water on reefs and coral rubble in perfect concealment because of their colouring and unfishlike appearance; both have a series of short, very sharp spines in the fin on the back. These spines have a shallow channel along their sides which connects to large grey-brown venom glands one each side. Anyone unfortunate enough to stand on a concealed stonefish receives puncture wounds which are automatically flooded with venom from the glands. The spines are sharp enough to penetrate a rubber soled plimsoll and only the stoutest boots (on a tropical reef!) really give safety.

Stonefish wounds are agonisingly painful. Frequently secondary infections set in which result in the loss of the limb, and deaths following a sting have been reported.

The stonefish is not, of course, known in the Atlantic but there is nevertheless a venomous fish commonly found in British seas. This is the weever (*Trachinus vipera*) which lives buried in sand in shallow water. A relatively small fish, it grows to about 5 in. and is very distinctive with its small, jet black dorsal fin with sharp spines. Both these spines and those on the head have venom glands, and a sting is extremely painful. The weever is much feared by the British east coast shrimp fishermen, for it lives in the same habitat as the brown shrimps, and occasionally bathers are stung.

Surgeon fishes or tangs are common in tropical waters, especially in the Indo-Pacific; they are colourful, deep-bodied grazers on algae growing on reefs and rocks. The name surgeon is applied because they have a scalpel-like spine on each side of the tail. These spines can be erected at will, pointing outwards and forwards and form a formidable weapon when the tail is moved from side to side and the fish is swimming ahead. Fortunately, other than accidently they rarely brush against human swimmers, but they use their spines freely in aggressive encounters with related species or other fishes on the reef.

Another fish weapon, which seems thoroughly offensive to any human encountering it, is electricity. Many kinds of fish have developed weak electrical power, perhaps as a means of signalling to one another (as in the rays), or of navigating in murky waters (as in the African freshwater fishes (*Gymnarchus* and *Mormyrus*), but some have developed powerful electrical properties by means of which they catch their food and fend off enemies. One of the best-known in freshwater is the electric eel, a native of north-eastern South America. It is long-bodied and eel-like, although it is not related to the eels, with a broad head, wide mouth and minute eyes. Most of the thick body is occupied by the electric organs, composed of columns of wafer-thin electroplates arranged in rows. These electroplates are connected in series, and although each produces only a small charge individually, combined they add up to a heavy shock. An electric eel in good condition can produce 500 volts and at times much more.

It lives in turbid water of creeks, streams and pools and from the size of its eyes it clearly does not see well in such conditions. It produces low-level electrical impulses more or less continuously, surrounding itself with an electrical field by means of which it can detect the movements of prey or predator. Large electric eels feed almost entirely on fishes and it is assumed are able to detect them in this way and stun them by means of a more powerful shock. On the occasions when a wading human comes in contact with a large electric eel he is likely to receive a considerable shock, sufficient it is said to knock a man down—so clearly its electrical powers have considerable protective value.

Electric rays (family Torpedinidae) are found in all tropical and warm temperate seas. Two species are found in British waters of which the more common *Torpedo nobiliana* is also the largest known, growing to 6ft in length and a weight of over 100lbs. They use their electrical batteries, which are housed in the thick wing-like pectoral lobes of the disc, to catch fishes, springing up from the bottom and wrapping the disc around the prey while instantly delivering a strong shock. The stunned, or dead, prey can then be eaten slowly, for all electric rays have a disproportionately small mouth. That it is an effective means of catching food is shown by the size, and number of fish, often actively swimming types, which they have been found to contain. The voltage produced by the common British electric ray has been measured

This two-winged flying fish displays the typically enlarged pectoral fins of the family. These are not in themselves propulsive units but they allow the fish to glide at speeds of up to 15mph over considerable distances. (One species has been known to achieve a 500ft glide.) Forward propulsion is gained by powerful strokes of the tail as the fish leaves the water, and during a glide a fish may 'touch down' and with a quick movement of the tail in the water become immediately airborne again. By this means these slender fish escape their predators. Flying fish are found in the tropics, both in the sea and in fresh water

at between 170 to 220 volts, so the shock can be severe. Fortunately, this species lives in moderately deep water (30ft and more) and is therefore not encountered by bathers or paddlers, but sea anglers and divers occasionally catch it, with painful results once the fish makes contact with its captor and the latter is earthed!

Mimicry and protective resemblances

While many fishes are well protected by their colouration and markings which merge into the background, a few seem to have become expert at resembling some other animal or plant. A protective resemblance is when an animal has the shape and colouration of, and sometimes behaves as if it were, an object of no interest to a predator. Mimicry, on the other hand, is more complex and occurs when one animal copies the form and behaviour of a known noxious or harmful, or in some cases beneficial, species. Observations on the latter have multiplied rapidly since biologists took to diving and observing fishes under natural conditions.

One of the best documented cases of mimicry concerns the cleaner wrasse *Labroides dimidiatus*, and the blenny *Aspidontus taeniatus*. Both are widely distributed in the Indo-Pacific in shallow water. The wrasse is well-known as a picker of parasites from other reef fishes and is brightly and boldly coloured as if to advertise its presence and to attract larger fishes to its station to be cleaned. The blenny is almost identical in colour and markings, most particularly in having a dark broadening stripe from snout tip to tail. It also swims in the hesitant, slow manner of most wrasses, certainly nothing like the rapid darting of a typical blenny. As the wrasse appears not only to be unmolested but even allowed to approach closely to larger fishes, the blenny clearly derives some protection from its resemblance to the wrasse. However, the relationship is more complex than it appears. The blenny has two massive, forward-pointing fangs in its lower jaw, and studies of its food have shown that it feeds on the fins and scales of other fishes. It seems then that by its mimicry of the harmless, and even useful wrasse, the blenny is tolerated by predators and also is allowed to approach closely enough to other fishes to make a sudden attack on them.

A similar mimetic relationship has been observed between a West Indian wrasse *Thalassoma bifasciatum* and a blenny *Hemiemblemaria simulus*. Rather more complex is the relationships between three blenny species found in the Red Sea, often in close proximity to one another. One of them, *Meiacanthus nigrolineatus*, has large fangs in its jaws with venom glands attached. Its bite can be regarded as painful and it is generally left severely alone by predatory fishes in the area. *Ecsenius gravieri* is very similar to it in both appearance and behaviour and, although not capable of inflicting a noxious bite, gains protection from other fishes by its mimicry. Both feed on small invertebrates. The third species, *Plagiotremus townsendi*, is a fin and scale biter of larger fishes, and again there is a strong resemblance to the other two species of the complex. The situation can be explained by supposing that the first species is the model which the other two mimic, deriving protection on account of its toxic bite. The third species is also a mimic of the second, and is able to approach fish prey closely on account of the harmless life-style of the second.

Not all mimetic relationships involve fishes alone. An Indo-Pacific snake-eel (*Myrichthys colubrinus*) is boldly marked with black bands around the body, exactly as in the venomous sea-snake *Platurus colubrinus*; it also swims in the same undulating manner. Biologists who have observed the two alive find them impossible to distinguish from one another without capturing them first, and sea-snakes are amongst the most deadly snakes known.

The batfishes of the genus *Platax* seem to specialize in deceptive similarities. The adults are deep-bodied, almost plate-like, with very long, well-developed fins, and because of their shape few except the largest predators could attack them. The young are more vulnerable and have resorted to various stratagems to avoid detection. The young of one species (*P. orbicularis*) float on their sides near the surface alongside floating yellowish leaves of *Hibiscus*, from which they are almost indistinguishable. In another species (*P. pinnatus*) the young are coloured black with a bright orange strip running around the outline. They swim on their sides gently undulating their fins, in a manner totally different to the behaviour of the adult fish. There seems to be no doubt that the young fish is mimicking a noxious invertebrate, probably a flat worm, some of which have a broad orange margin against a dark background, or possibly a nudibranch mollusc, of which a few are similarly coloured. Both animals produce noxious secretions, and some nudibranchs make second-hand use of coelenterate stinging cells; their bright hues and patterns are obviously warnings to potential predators.

Very many marine fishes appear to make an effort to look like vegetation. Many pipefishes strongly resemble in colour, and even shape, the seaweeds or eel grass amongst which they live. The young of the leatherjacket (*Oligoplites saurus*) is commonly found in the coastal waters of Florida pretending to be a yellowing, decaying mangrove leaf floating at the surface, while in the same area young spadefishes mimic the black over-ripe seed pods of the red mangrove. The best known of leaf-mimics is, however, not a sea fish but the freshwater leaffish (*Monocirrhus polyacanthus*) of the Amazon and Rio Negro river systems of South America. Its body is deeply compressed, and the tail end is curiously foreshortened while the snout is long with a huge mouth, and at the tip of the lower jaw a stout barbel. It grows to about 4in. in length. Its colouration is variable, but is usually some shade of brown, mottled with green and grey. It drifts in the surface water in a head-down posture, often at an odd angle, and looks strikingly like a dead leaf. Even the chin-barbel is involved, for this is held stiffly forward and adds to the deception by resembling the leaf stalk!

While the leaffish may derive some protection from its excellent imitation of a leaf, the prime purpose of the disguise seems to be the capture of food. The leaffish eats fishes exclusively, and apparently drifts towards its prey, propelled by movements of the colourless pectoral fins, until it is within striking distance, when the capacious mouth swings open and forwards and its prey is engulfed. Relatives of the leaffish are found in Africa and Asia, and although they mostly feed on fishes and are adept at stalking their prey, none have refined the art of deception to the degree of the South American species.

Cleaner fishes

Cleaning relationships amongst land animals have long been recognized; tick birds and rhinoceroses, egrets and cattle, are well known for their relationship, the bird in each case feeding off the insects and parasites attracted to the mammal. Similar relationships exist in the sea, and in fact seem to be more numerous there, although many of them are only recently recognized and no doubt others remain to be discovered.

Amongst fishes the family which contains the most numerous cleaners is the wrasse family—Labridae. The best-known cleaner is undoubtedly the blue streak (*Labroides dimidiatus*), a 4in. long inhabitant of the tropical Indo-Pacific. It is extremely abundant on reefs where it adopts a territory in which it displays in eye-catching manner, conspicuously flashing its brilliant blue back contrasted with a black length-wise stripe. Larger fishes come to these cleaning stations, which are usually clearly visible by being on the open side of the reef, and while they hover in the water, their fins fully expanded and only lightly moving, the blue streak cleans them of external parasites, as well as picking wounds and infections clean. Often the host will open its mouth and gill covers, fully permitting the cleaner to investigate these frequent sites of infestation by parasites, and one observer has reported seeing a 2in. long *Labroides* enter the opened mouth of a 4ft long moray eel which gaped its jaws obligingly.

The blue streak has also been seen to clean other wrasses, amberjacks, groupers, parrotfishes, red mullets and damselfishes, and obviously performs a useful service in keeping these fishes free of parasites. So obsessed is it with cleanliness that one diving biologist has reported a *Labroides* attempting to pick the hairs off his legs!

Many other species of fish clean other fishes. In Californian waters the small brownish wrasse known as the señorita has been seen to clean a number of larger fish, and examination of their stomach contents has revealed parasites and pieces of infected skin. In the Caribbean other wrasses, young butterfly-fishes and several species of goby (*Elecatinus* spp.) have been reported to clean other fish. For long it was thought that these cleaning symbioses were entirely confined to tropical waters. However, this now seems to be because more underwater observations are made in the tropics than in the uncomfortably cool, often clouded waters of temperate seas. Recently, cleaning behaviour has been reported in New Zealand waters for a small species of wrasse, and even in British seas the rather uncommon small-mouthed wrasse has been found to feed on fish parasites. Most, if not all, of these cleaner fish are distinguished by bright colouration and bold markings, which seem likely to serve as an advertisement for the cleaner fish ('guild-mark' has been adopted as a descriptive term for it), much as many barbers' shops advertised their presence by a spirally coloured pole.

It had been claimed that the work of the cleaner fish was essential to the continued health and even presence of fish in the vicinity. This stemmed from an experiment in which all the cleaners were removed from two small isolated reefs in the Bahamas, and from which all the territorial fishes disappeared within two weeks. More recently, similar experiments in the Pacific Ocean where all the blue streaks were captured on a reef have proved inconclusive.

Cleaning relationships have similarly been proved between some of the very strange remoras, or suckerfish, and their hosts, and have thus explained a previously insoluble mystery. The suckerfish family consists of generally slender-bodied marine fishes with a unique sucker on the head and back, oval in shape and slatted like a venetian blind with a raised rim around its circumference. With this sucker they cling to the underside of a shark, other large fish, or even whale. For long it was thought that the suckerfish merely attached itself for the ride, and fed on the scraps left over from the shark's meals, but most sharks when feeding do not leave many scraps, most often the meal is bolted whole! The discovery that *Remora remora*, which is the species mostly found on sharks, frequently had external parasitic crustaceans in its stomach first led to the suspicion that the relationship was that of parasite-picker and host. Since then it has been found that other suckerfishes, including *Remoropsis brachypterus*, which is found on marlin, tunny and the swordfish, also feeds on parasites. Suckerfishes are also found in the gill cavity of the giant manta rays, and one species (*Remiligia australis*) lives on various whales; it would not be very surprising if these too were in time proved to be parasite pickers.

Fish-life in the deep sea

In the early days of marine biology it was believed that below the sunlit zones of the sea there existed a lifeless zone, the so-called Azoic zone. However, a century ago, during the voyage of HMS *Challenger*, the first oceanic expedition to be mounted on a grand scale, this theory was disproved and animals, most of them new to science, were dredged and trawled from the deep sea.

There is something fascinating in the capture of fishes in the great depths of the oceans. The deepest known capture is of a brotulid fish (*Bassogigas*) in the Puerto Rico Trench at a depth of over 25,000ft, nearly five miles. Fishes of this genus have been found in several of the great trenches of both the Atlantic and Indian Oceans, and clearly they have specialized in deep sea life. Surprisingly in some ways they have a large and functional swimbladder, and it has been suggested that the oxygen stored in it may be used for respiration in areas where dissolved oxygen in the water is low. If this is so it would be yet another example of the fascinating adaptations by fishes for life in difficult circumstances.

In the deep sea the density of animal life is lower than in the shore or surface regions of the sea and many of the adaptations are in response to this sparsity of life. In the absence of plant life all animals are predatory, and most of the deep sea fishes have huge teeth and enormous jaws. It seems that opportunities to feed are few, and no chance can be missed even if the prey is nearly as big as its captor, and occasionally fish are caught with prey in the gut as long as they are themselves! The viperfishes (*Chauliodus* spp.), which are found in all deep seas are a perfect example. The fangs are so huge that the jaws cannot be shut tight, but the latter are loosely hinged so that they can swing wide open. The skull is also loosely attached to the spinal column so that the head can be swung upwards when the jaw drops down. In doing so the vulnerable gills, and the heart and major blood vessels are moved away from the gullet, so that even if a large prey is attacked and swallowed its dying struggles will not damage any of these delicate and important organs.

The viperfish has a long curved ray with a lighted tip at the front of its dorsal fin which is dangled enticingly before its snout to attract prey. Many other deep sea fishes have luminous lures on the snout which serve the same function. These are probably best developed in the deep sea anglerfishes, which all have elaborate fishing rods and lighted lures on their snouts. Angler-

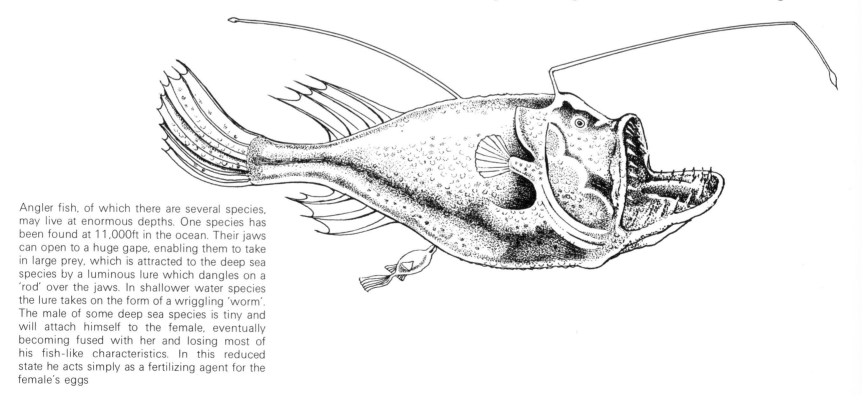

Angler fish, of which there are several species, may live at enormous depths. One species has been found at 11,000ft in the ocean. Their jaws can open to a huge gape, enabling them to take in large prey, which is attracted to the deep sea species by a luminous lure which dangles on a 'rod' over the jaws. In shallower water species the lure takes on the form of a wriggling 'worm'. The male of some deep sea species is tiny and will attach himself to the female, eventually becoming fused with her and losing most of his fish-like characteristics. In this reduced state he acts simply as a fertilizing agent for the female's eggs

fishes of the family Ceratiidae have an even more fascinating adaptation to the vastness of the deep sea. The females are often large (some species grow to 24in. in length), but once the male attains an inch or so it attaches itself to the female and becomes parasitic upon her. The blood supply of the two fishes is joined and the male is supplied with such oxygen and nourishment as are necessary. As most of its organs become smaller with disuse it probably makes a minimal demand in this way, but when spawning takes place there is no question of failure to find a member of the opposite sex; fertilization of the eggs is ensured by the involuntary presence of the male.

Ceratoid anglerfishes also possess other interesting features. Free swimming males have larger nasal organs and better-developed olfactory lobes in the brain than do the females of the same species. There is no doubt that their enhanced sense of smell enables the males to find the females in the unlit sea. After they become parasitic their nostrils degenerate.

Light organs are well-developed amongst deep sea fishes, although they are most specialized in those fish which live in the upper mid-waters of the ocean. Lanternfishes, of which there are more than two hundred kinds in all the large seas and oceans, have a pattern of rounded pearly light-organs along their sides. Each species has its own characteristic pattern, so there is no doubt that schools of one species can keep together and recognize one another. However, some species have an enormous light organ in front of each eye; do they use this as a kind of head light to see where they are going and to illuminate their surroundings? Viperfishes and their relatives the stareaters (Astronesthidae) and scaly dragonfishes (*Stomias* spp), as well as other groups, have many more light organs than lanternfishes and when switched on might be expected to produce a scintillating display. Most of them also have a smaller single light organ beside each eye, and it has been suggested that this, presumably by being alight all the time, helps to adapt the eyes against dazzle when the main lights are working. This explanation seems to provide a good working hypothesis, but has yet to be proved, and the function of these eye lights is one of the many thousands of problems which remain to be solved amongst the fishes of the sunless seas.

Some species of fish, such as these herrings, form huge shoals. Though most relatives of herrings are marine fish, living near the surface, some are freshwater inhabitants. Apart from mass migrations to spawning grounds, herrings also make other seasonal migrations. A female may lay up to 100,000 eggs on the bed of the spawning grounds and the young fish eventually find their way to shallow water round the coasts. Small herrings constitute the well-known delicacy whitebait

Above The young of African mouthbrooders are protected from danger within their mother's mouth. Here the mother is spitting out a mouthful of her brood, retrieved from the edge of the territory. In extreme danger, or by mistake, the mother may actually eat her own children

Left The coral fishes of the Great Barrier Reef off the north-east coast of Australia are unusually brilliantly coloured. Despite their brilliance, however, they are naturally camouflaged among the bewildering variety of colours found in coral reefs. This yellow-faced anemone fish lives among the tentacles of sea anemones, apparently immune to their stinging cells. It is thus also immune from most enemies

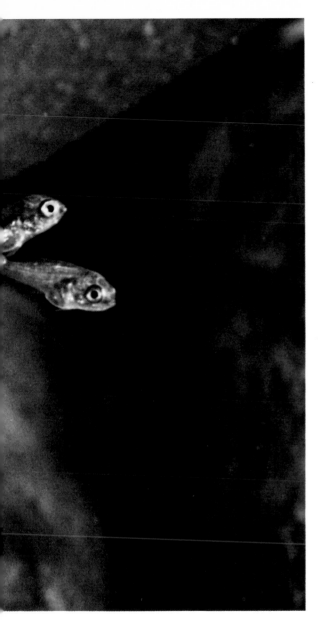

Above Pike eats pike. These voracious fresh-water fish are found in Europe and North America. They grow to nearly 5ft and will eat anything within reach, including smaller pike and other fish, birds and small mammals

Right Remoras have sucker discs on their heads by which they can attach themselves to larger fish such as sharks—also turtles. While they get a free ride, they also clean parasites off their hosts and will detach themselves for the pickings of a large kill

Overleaf The white shark, one of the most fear-some killers in the sea, though not as big as the whale shark (up to 60ft), may reach a length of 22ft and weigh over half a ton. It will tackle anything, even other sharks. The white shark pictured has ripped out a huge chunk of flesh from its victim. Its external gill slits, typical of sharks, can be clearly seen

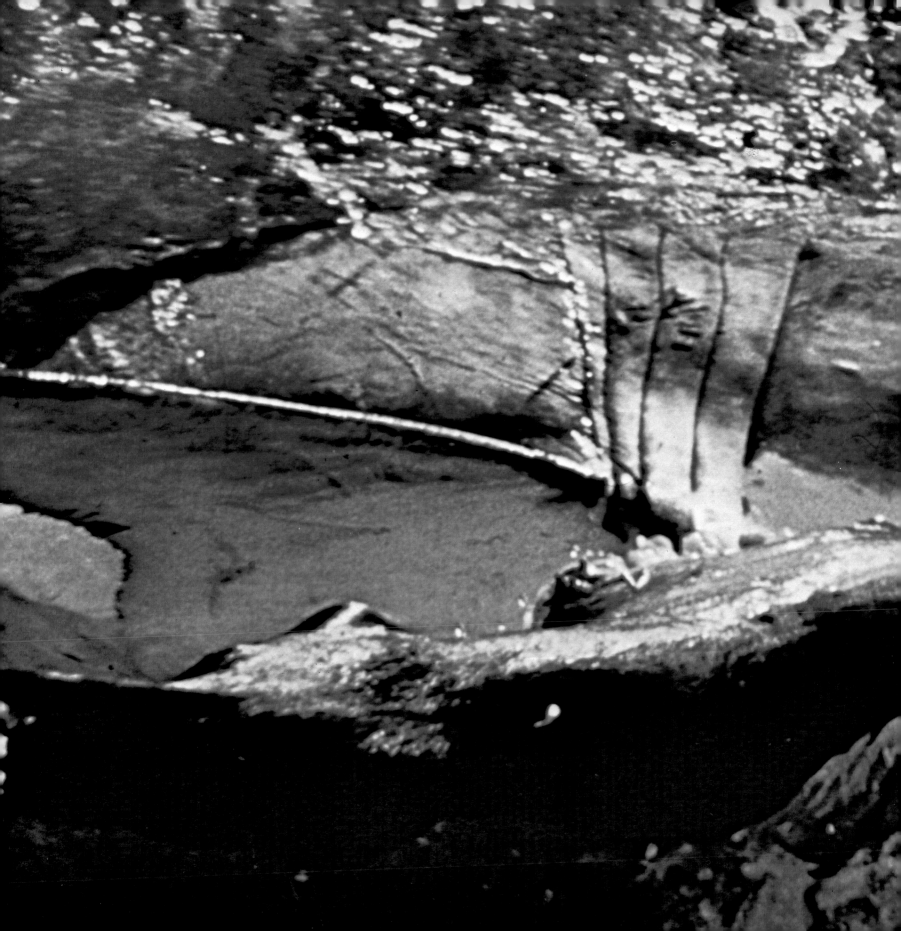

Left Salmon have a complicated and migratory life history. Here a cock sockeye salmon from British Columbia keeps guard while the hen digs a nest in the gravel bed. The eggs (*above*) hatch and the tiny salmon, their yolk sacs still attached, begin their long development through the fry stage (*below left*) until eventually, as mature adults, they in turn will spawn. After their exhausting journey to the spawning grounds the adults usually die (*below right*)

Insects

by Michael Chinery

Introduction to insects

Unlike the animals discussed earlier in this book, the insects are invertebrates, which means that they have no backbone. In fact, they have no internal skeleton at all and their body muscles are anchored to the tough outer coat. The insects belong to the arthropods, a very large group which they share with the crabs, woodlice, spiders, scorpions, centipedes, and various other creatures. Most of these animals have four pairs of legs or more, but the adult insect has only three pairs of legs and a single pair of antennae or feelers. Most adult insects also have two pairs of wings, although these are not very obvious in beetles and some other insects whose front wings form protective covers for the body. The wings of some butterflies and moths span about 30cm. (12in.), but the majority of insect wings are much smaller. The flight mechanisms also vary a good deal, especially in the rate of wing flapping. Butterflies get plenty of lift from their large wings and they can fly quite happily with an average of about ten wing beats per second. Large moths, whose bodies are generally heavier than those of butterflies, flap their wings about sixty times per second in order to remain aloft. Small-winged insects have to work much faster than this, and many small flies have wing-beats exceeding 200 per second. Some mosquitoes vibrate their wings about 600 times per second, giving rise to the familiar hum or whine which tells us that we are being attacked. The thorax of these small insects has a very special elastic construction which enables the wings to beat at such high rates.

The ability to fly has been very important in the evolution of insects, for it has endowed them with tremendous mobility—mobility to find food and to escape from their enemies, mobility to find their mates, and mobility simply to spread out into new pastures. These are some of the factors which have made the insects such successful animals. There are, nevertheless, many flightless insects in the world today. The most primitive insects, including the little silverfish of our kitchens, never have had wings, but others have actually lost their wings during their evolution. Just as the kiwi and various other island-dwelling birds have found flight unnecessary in areas free from predators, so some insects have found niches in which they do not need to fly. The blood-sucking fleas, for example, spend their days wandering through the fur or feathers of warm-blooded animals, where wings would be not merely unnecessary but a positive hindrance. The processes of natural selection have therefore gradually removed the fleas' wings by favouring, generation after generation, those individuals with the smallest wings.

Like most hawk moths, the elephant hawk moth possesses an immensely long tongue (proboscis) with which it can reach nectar deep in the corollas of flowers. Fast fliers, when feeding they hover in front of their food plant, their wings beating more than 50 times per second. This species gets its popular name from the larva, which somewhat resembles an elephant's trunk

The flies are all liquid-feeders, but they have several different types of feeding apparatus. The house-fly, for example, has a spongy 'mop' with which it soaks up liquids of various kinds, while the blood-sucking mosquito has mouth-parts in the form of minute needles fitted together to form a hypodermic syringe. Although often used against us, sometimes with painful results, the construction of this syringe is really something to be admired. It is finer than a human hair, and yet it contains two distinct channels—one to pump anti-coagulant saliva into the wound, and the other to suck up the blood. The whole apparatus is kept in a sheath when not in use and the sensitive tip of the sheath is used to pick out a suitable feeding site. Only female mosquitoes actually suck blood: the males are content to suck nectar. It seems that the females need a feed of blood before they can lay their eggs.

The aphids and the other bugs, all of which feed on the juices of plants or other animals, have piercing mouth-parts quite similar to those of the mosquitoes, but the nectar-sipping bees have a very different set of 'tools'. The 'lower lip' is drawn out into a blade-like tongue whose edges curl round to form a channel for the nectar. The length of the tongue varies from species to species, the longer-tongued species being able to get nectar from some quite deep-throated flowers. Unlike the butterflies and moths, the bees retain their biting jaws and use them for building. Some of the short-tongued bees also use their jaws to bite through the bases of tubular flowers and then 'steal' nectar which they cannot reach in the normal way.

All insects begin their lives as eggs, although the eggs of certain aphids and other insects hatch while still in the mothers' bodies. The females then give birth to active young, and some aphids have three babies every day for about ten days. Such a birth rate seems very high by human standards, but it is really rather low when compared with the output of most egg-laying insects. The prize goes to the queen termite, who might lay more than 10,000 eggs in a single day and produce many millions during her lifetime.

The young insect hatching from the egg may look like a small and wingless version of the parent insect. Youngsters of this kind are called nymphs and they are found among the dragonflies, grasshoppers, bugs, and the other more primitive groups of insects. The eggs of butterflies, bees, beetles, and flies give rise to another type of youngster. This bears no resemblance at all to the adult insect and it is called a larva. Both kinds of youngster have to grow in stages because they have a tough outer coat or cuticle which stretches little. After a few days of almost continuous eating, the young insect becomes too large for its coat and it has to moult. This is a somewhat hazardous process for the insect and it takes a day or two to complete. Like most living processes, moulting is controlled by hormones. In some insects, and possibly in all of them, the brain detects a feeling of tightness when the old coat is getting too small and it releases a hormone into the body. This hormone affects glands in the head or the thorax and it causes them to release another hormone which affects the skin and starts the moulting process. The insect is

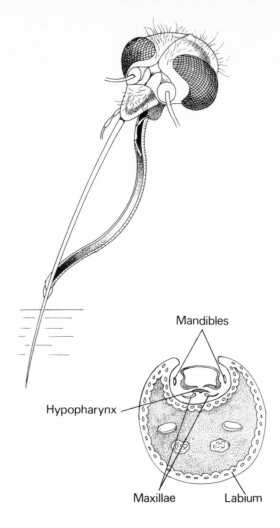

Mandibles

Hypopharynx

Maxillae   Labium

The mouthparts of the adult female mosquito consist of minute needles fitted together to form a hypodermic syringe. The syringe is finer than a human hair and contains two distinct channels: one pumps anti-coagulant saliva into the puncture to prevent the blood from clotting, while the other is used as a straw to suck blood up into the insect. When not being used, the whole apparatus is kept in a sheath. Most female mosquitoes must have a meal of blood before they can lay eggs. Males do not suck blood but feed on nectar

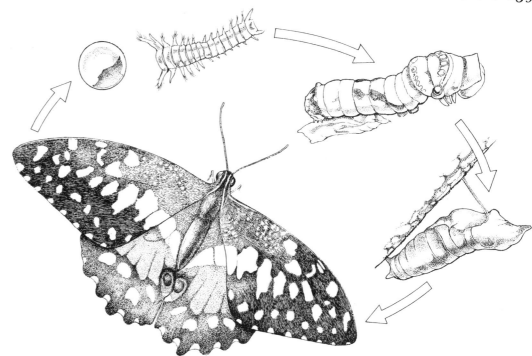

This swallowtail butterfly shows the typical life cycle of the majority of insects. It undergoes a complete metamorphosis in four stages. When the larvae hatch from the eggs they have three pairs of legs, a pair on each of the first three segments, and five pairs of pro-legs further back. The latter legs are not present in the adult. The larva eats and grows, moulting several times in the process. When fully grown, it pupates, forming a hard shell round its body. Inside the pupa the body is almost totally broken down and built up again into its final adult form. At this point the adult breaks the shell of the pupa and struggles out. After a period of rest while the wings expand and waste products from the metamorphosis are pumped out of the body, the fully winged adult butterfly, or imago, flies. Eventually it will mate and the female will lay the eggs that start the cycle off again.

resting at this stage and, under the influence of the moulting hormone, the cuticle becomes detached from the underlying skin. Useful materials are withdrawn from the cuticle and the skin produces a much-folded new cuticle underneath the old one. The latter is now very thin and brittle and the insect breaks out of it, often by puffing itself up with air or water. The new loose-fitting cuticle allows the insect to begin a further period of growth, but it gradually gets tighter as its folds are taken up and the whole moulting process has to be repeated before long. Most insects undergo four or five moults during their lives, but some species undergo considerably more than this.

Whether the young insect is a nymph or a larva, it has to undergo a certain amount of change or metamorphosis before it becomes adult. A nymph gets more and more like the adult at each moult, the most noticeable development being the appearance and growth of the wings. These get larger at each moult and become fully developed when the insect moults for the last time and assumes the adult form. The metamorphosis is controlled by a hormone produced in a gland just behind the brain. It is known as juvenile hormone and its presence in any quantity causes the moulting insect to remain in the immature state. Experimental removal of the gland causes young nymphs to turn into very small adults, while injecting the hormone into fully grown nymphs causes them to turn into even larger nymphs instead of into adults.

A larva follows a rather different path of development and the change to the adult form takes place all at once when the larva has reached its full size. At each of the early moults, the larva merely gets larger, but when it is fully grown it moults into a pupa or chrysalis. This is popularly regarded as a

resting stage, because pupae do not feed and rarely move about, but there is tremendous activity inside the pupa. Under the control of various hormones, the whole larval body is broken down and re-built into the body of the adult insect. The conversion is often remarkably quick—only a few days in some flies—although the adult insect may not actually emerge from the pupa for several months.

The bulkiest insect, the goliath beetle of equatorial Africa, is only about the size of a man's fist and it weighs no more than about 30 grams (1oz.). The great majority of insects are very much smaller than this, however, and most of them could sit very comfortably in a teaspoon. But what they lack in size, they make up with their immense numbers. Some years ago it was estimated that an acre of meadowland in England supported something like 400 million insects. Nearly all of these were, of course, minute soil-dwelling creatures, but they were insects nevertheless. If we take this figure as an average population density for the whole land surface—there are very few insects in the sea—we arrive at a world population of something like 12,000,000,000,000,000,000 insects. Such a figure can, of course, be only an estimate, and the true population may well be ten times larger or ten times smaller, but it illustrates very clearly that there are enormous numbers of insects in the world. This vast population is divided among about a million known species and probably another million species not yet discovered.

The ability to fly has already been mentioned as an important factor in the success of the insects. Their small size has also played a big part in this success, for it has enabled them to colonize places and use food materials unavailable to larger animals. Many insects, for example, are able to pass almost their entire lives in the narrow spaces between the upper and lower surfaces of leaves. A third factor of immense importance has been the amazing ability of the insects to adapt their feeding apparatus to make use of every available source of food, whether it be liquid or solid, living or dead. The primitive insects had mouth-parts of a simple biting type, with hard jaws used for biting and scraping. Several insect groups, including the grasshoppers and the beetles, have retained this kind of apparatus to the present day and use it to feed on a variety of solid foods. Biting jaws are no good for dealing with a liquid diet, however, and the mouth-parts have undergone considerable modification in the various groups of liquid-feeding insects. Butterflies and moths feed mainly on nectar, and their mouth-parts are in the form of a slender 'drinking straw' called the proboscis. One hawkmoth from tropical America has a proboscis about 25cm. (10in.) long— much longer than the rest of the body. But not all butterflies and moths possess proboscises. Many moths, in common with various other insects, do not feed when they become adults. They take in all their nourishment while they are caterpillars and spend their short adult lives burning it up in the all-important business of mating and reproducing their kind.

The body of the adult house-fly is covered with a horny skeleton, to which the muscles are attached. To grow, the insect must shed this 'exoskeleton' by moulting. The fly has a head, a thorax and an abdomen. The segmented legs spring in three pairs from the three segments of the thorax, and the second and third segments also bear the veined wings, though in the case of the house-fly the second pair of wings has been reduced to two stubs, called halteres. Air is breathed through a network of tubes (tracheae) via air holes (spiracles) along the body. The sexual organs are located in the last segments of the abdomen

Insect behaviour

Studying insect behaviour, just like studying the behaviour of other animals, necessitates a thorough understanding of the sensory capacities of the creatures. In other words, we must know just what they are or are not able to detect in the way of stimuli from their surroundings. But knowing the capabilities of an insect is only one of the requirements. The investigator must discover which stimuli actually trigger off responses in the insect, and not merely find out what signals the animal can detect. Like us, the insects cut out or reject a lot of the information with which they are bombarded, and they react only to signals which are important to them *at the time*. Research along these lines faces problems right from the start because the insects' sense organs are so different from our own, but the patient work of the Austrian biologist Karl von Frisch and many other scientists has shown that these problems can be overcome. After thousands of experiments and observations von Frisch was able to prove that honey bees have good colour vision, although it is not quite like ours in that the bees cannot see reds but they can see ultra-violet light. This means that they do not necessarily see flowers and other objects in the same way that we do, and it would be easy to reach false conclusions if we lost sight of this fact.

In spite of the many difficulties, we now know a great deal about insect behaviour—not just what they do, but why and how they do it. The bulk of insect behaviour is innate or instinctive, carried out according to inbuilt patterns which are just as much a part of the animal as its shape and colour. These patterns ensure that the insect does the right thing at the right time, each piece of behaviour being triggered off by a particular signal or releaser. But research has shown that the insects are not quite the automatons that they were once thought to be: many of them are able to learn and, even if the innate behaviour is never modified very much, the behaviour is at least flexible. It usually happens that the learned behaviour is superimposed on the innate pattern. Good examples of this are afforded by some of the solitary wasps, whose basic behaviour pattern involves constructing a nest chamber, filling it with one or more paralysed insects or spiders laying eggs in it, and sealing it up. The completion of each phase normally triggers off the next stage, and most of the insects are unable to go back and repair any damage. In this respect, their behaviour is rather stereotyped, but some species actually have several nests on the go at once. Each nest may be at a different stage of completion and the insect must obviously learn not only the positions of the nests but the stage reached by each one. And it must remember these things.

Most insects have some ability to learn, but learning is best developed among the bees, the wasps, and the ants. Under natural conditions, they pick up most of their information by the process known as latent learning, in which information about the surroundings is absorbed while the insects are going about their normal business. The solitary wasps again provide a good example. Dr Niko Tinbergen, famous for many studies of animal behaviour, demonstrated it very well by surrounding the burrow of one of these wasps with a ring of pine cones. The cones were obviously seen every time the wasp entered or left the nest, and their existence was absorbed into the insect's memory. The wasp then merely homed in on one of the cones: it had learned that its nest was in the centre of the ring. Unfortunately, it could not undo its learning and it was unable to find the nest again when the ring of cones was moved to one side. Honey bees use the same system of latent learning to remember the landmarks round their hives.

Some insects can also make use of associative learning. Honey bees, for example, soon learn to associate certain colours with food if they are provided with syrup in dishes of those colours. They will then come to the colours if no food is present. The bees which bite through the bases of flowers to 'steal' nectar are also using the system, associating this behaviour with the reward of food. Ants have particularly good powers of learning and they have no trouble in mastering a six-point maze if they are rewarded at the end. In fact, they take only about three times as long to master the maze as a rat, and rats are widely regarded as fairly intelligent mammals. The insects are, however, incapable of insight learning: they cannot reason or think ahead, nor can they really solve problems, although some experiments have shown that solitary wasps can sometimes find their way round glass screens that are put in front of their nests.

Before going on to look at some of the fascinating examples of insect behaviour, it is important to point out that, without the ability to think ahead, the insects cannot really carry out any purposeful actions. Although we often say that insects or other animals do something *in order to* achieve a certain result, we must remember that this is a very anthropomorphic way of looking at animal behaviour and that the animals do not act in this way, however purposeful their behaviour might seem to us. They are merely following instincts which have evolved *because* they achieve the correct result. Some biologists get rather hot under the collar if *any* human thought or emotion is read into the behaviour of animals, and we must certainly avoid using words like *pretend*: leaf-insects in no way pretend to be leaves—they merely resemble them. On the other hand, there is no harm in saying that an insect enjoys a good feed, as long as we do not visualize a picture of the insect sitting there and thinking 'what a nice meal this is'.

Honey bee language

As a result of its usefulness to man and its complex social behaviour, the honey bee can justly claim to be the most deeply studied of all insects. Scientists have discovered how it keeps its hive warm in winter and cool in summer, how it changes jobs as it gets older, and how the queen controls the behaviour of the whole colony—perhaps 50,000 bees at the height of summer—by secreting a constant flow of 'queen substance' which is licked

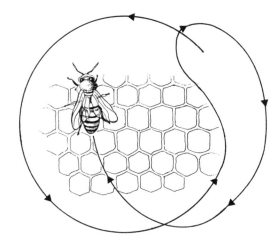

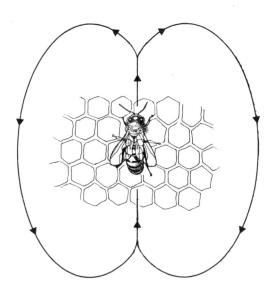

The remarkable dances of the honey bee take place on the vertical surfaces of the combs. Their purpose is to signal to the other workers in the hive the presence of nectar sources. A returning worker performs a simple round dance or a more complex figure-of-eight dance on the comb. The angle between the axis of the dance and the perpendicular indicates to the other workers the direction of the source of nectar relative to the position of the sun. (Bees can indicate this direction even on overcast days.) The distance, quantity and type of nectar are also transmitted to the hive. A simple, brisk round dance (above) means that the food is at up to 50yds away. The lively figure-of-eight dance below, with little tail waggling, shows food at about 100yds. At longer distances the bees' abdomens waggle fast on the cross run of the figure-of-eight though the outer circles are slow

from her body and passed round to all the other bees. But the most amazing discoveries have concerned the bees' language. The bees' ability to 'talk' to each other and to tell each other where to go for a good meal was first explained by von Frisch in the 1920s. Using glass-walled hives, he saw that bees returning from the fields 'danced' on the combs and that other bees followed the dances before leaving to look for food. Years of careful experimentation and observation led von Frisch to the conclusion that the form of the dance tells the other bees exactly where to find a good supply of food. This seemed so incredible at the time that most other biologists refused to believe it, but when they looked for themselves they saw that von Frisch was right. It is now fully accepted that the bees do 'talk' to each other with their dances, although some scientists still have reservations on just how some of the information is passed on.

There are actually two main forms of dance—the round dance and the waggle dance. The round dance merely involves the returning bees in running round in circles, first this way and then that, and it indicates that they have found worthwhile sources of nectar within about 50 metres (50yds) of the hive. Other worker bees get excited and they leave the hive in all directions to search for the food. The more elaborate waggle dance is performed when the food is further away and it gives precise directions. It consists of a rather squashed figure-of-eight movement with a 'straight run' through the centre. The direction of the straight run is adjusted so that it makes an angle with the vertical which is exactly the same as the angle between the sun and the food. The following bees have no trouble in detecting the angle and they fly out of the hive in just the right direction.

The waggle dance also indicates the distance of the food from the hive, but the actual method by which the distance is indicated is not so clear. The area covered by the dance is greater when the food is further away, and so the number of loops and straight runs covered in a given time is reduced. The bee draws particular attention to herself during the straight run by quivering her abdomen, and it is generally thought that the length of the straight run or the number of runs per minute is a measure of the distance between the hive and the food. Recent work, however, suggests that the following bees might actually be responding to sounds given out during the straight run and not to the length or frequency of the run itself.

The bees do not, of course, measure the distances in terms of metres or other units of length. They measure in terms of the effort needed to reach the food. If a food source is placed downwind of the hive a returning bee will actually indicate a shorter distance than one returning from a food source placed the same distance upwind. Bees flying out in response to the dances, however, will expend the indicated amount of energy and, helped or hindered by the wind, will arrive at the correct place. It is interesting to notice that, before flying out to an indicated food source, the bees seem to assess their own energy reserves and take on extra honey if necessary. The instructions given during the dance are precise enough, and the reactions of the following bees are accurate enough to guide them to within about 150 metres (150yds) of the food even if it is 5000 metres (3 miles) from the hive. The bees can then use their eyes and antennae to home in on the food. Dances are not usually performed by bees returning home from more than about 5000 metres (3 miles): there is no point in encouraging other bees to fly so far because they will burn up more honey than they can collect.

As well as an uncanny ability to measure angles, the bees have a wonderful sense of time which allows them to correct their flight paths to counteract the changing position of the sun. The existence of this time sense has been ably demonstrated by von Frisch and by his pupil Martin Lindauer. The latter trained a colony of bees to feed at one place just before sunset and at another place just after sunrise. This in itself was sufficient to show that the bees had a good sense of time, but Lindauer's experiment did not stop there. By shining a light into the hive, he was able to wake up the bees and get them to dance in the night. Before midnight the dances indicated the evening feeding place, but the bees changed over at midnight and started to indicate the morning feeding place.

For a final glimpse into the fascinating world of the honey bee, we can look at its dialects—yes, just as the human species has its various races and languages, so does the honey bee. The differences between the honey bee languages are less marked than those between human languages, but they are still sufficiently great for one race of bee to be puzzled by the language of another. They all perform round dances and waggle dances in the same way and they indicate directions in the same way, but they dance at different speeds. The bees of one race cannot therefore understand what bees of another race are saying about distance. This is not really any problem to the bees, of course: it is only when interfering bee-keepers mix the races in one hive that they ever meet the language barrier.

Farmers and herdsmen

Ant societies, like their human counterparts, can be roughly divided into hunting, herding, and farming communities. The hunting ants, in common with primitive human societies, wander freely in search of food and make temporary homes here and there as they go. The herders and farmers, however, have settled homes. The most famous farming communities are those of the leaf-cutting ants from tropical America. These ants cut pieces of leaf from various plants and they often destroy crops, but do not actually eat the leaves: they use them to make 'compost heaps' on which they grow special fungi. The fruiting bodies of these fungi are the ants' sole food and huge quantities of leaves are needed to maintain the crop. Processions of ants are continuously moving back to the nest, with each ant waving a piece of leaf above its head. This habit has earned them their alternative name of parasol ants. On entering the nest, the ants chew the leaf fragments thoroughly and then add them to the fungus beds. A young queen leaving for her marriage flight instinctively picks up a small piece of the fungus bed and takes it with her in a small pouch near her mouth. The right kind of fungus is thus automatically introduced if she manages to set up a new colony. She actually grows her first crop on some of her own crushed eggs and excrement and she does not start to rear her first worker brood until the fungus is ready for eating. The workers are of three main types, differing mainly in size. The large ones defend the nest, the middle-sized ones collect leaves, while the smallest ones cultivate the fungus beds. But the small ones have another fascinating job as well: they appear to act as body-guards for the leaf-collectors when they are out in the fields. The ants are plagued by a small parasitic fly which tries to lay its eggs on them. They are defenceless while they are cutting or carrying leaves and it seems that the smallest workers often go along to protect them. These little workers ride on the backs of their larger sisters, or else on the leaf fragments, and they use their snapping jaws to keep the flies at bay. This is the true spirit of social life: it does not benefit the small workers to go out—in fact, they would be much safer if they stayed in the nest—but their protective action is of considerable benefit to the colony and the species as a whole.

When it is known that some ants grow their own mushrooms, it may not come as such a surprise to learn that some also keep and milk their own cows. The 'cows' are, of course, aphids and the 'milk' that they yield is honeydew. This sugary secretion is a waste product as far as the aphids are concerned—a means of getting rid of the excess sugar that they take in with their constant diet of sap—but it is eagerly sought by the sugar-loving ants. Some ants merely lap up the honeydew whenever they find an aphid, but others literally take care of the aphids in a manner very similar to that in which we look after cattle. The aphids are carried from plant to plant and they are 'milked' regularly: the ants stroke their bodies and they respond with a flow of honeydew. Aphids from far away are often brought closer to the ants' nests, and the ants may even build shelters for them. These are usually in the form of earthen pipes built up around the stems on which the aphids are feeding. Some aphids are even carried into the ants' nests and installed in special chambers through which run roots on which the aphids can feed. During the autumn the ants may carry the aphids' eggs into the nest, and they then care for them until the spring. This last is a really incredible piece of behaviour because the aphids' eggs are of no value to the ants until they hatch. It looks very much like an intelligent action involving a high degree of knowledge and foresight by the ants, but we can be sure that it is an instinctive act that has evolved as part of the ants' innate behaviour pattern. Or can we?

Slave-making ants
The typical ant colony consists of one or more queens, who do nothing but

lay eggs, and a number of sterile females called workers. The latter do all the work of nest-building, food-gathering, and brood-rearing. Male ants appear only at certain times of the year when there are new queens to be fertilized. There are, however, many variations on this basic pattern. Some ant species, for example, have no workers. The queens lay their eggs in the nests of other ants and, rather like cuckoos, leave the other species to rear the youngsters. Several kinds of ants living in the northern hemisphere employ slaves to do some or all of their work. *Formica sanguinea*, the blood-red ant of Europe and North America, frequently raids the nests of related species and brings back a number of pupae. These pupae are installed in the *sanguinea* nest and, when the adults emerge, they start to work just as they would in their own homes. Slaves are not essential to the blood-red ant and many of its nests are self-supporting, but the amazon ants of the genus *Polyergus* are completely dependent upon slave labour. The amazon queen lays her first eggs in a small hole and rears a few small workers, but these are quite unable to build a nest or to rear any more larvae. They are, however, equipped with powerful jaws and they set off to plunder the nests of other species. Not until the stolen pupae produce workers does the amazon nest begin to take shape. More amazon workers are then reared, but they remain incapable of carrying out the domestic chores and they make regular slave raids, often in large armies, to ensure a continuous supply of builders and nurse-maids for their nests.

Weaver ants

The ants never make anything quite as elaborate as the paper or wax cells of the wasps and bees and their nests are remarkably simple when compared with the fine architecture of these other insects. The nest of the weaver ant is among the simplest of them all, being composed merely of a few over-lapping leaves fixed together to form a bag. But when we come to consider just how the leaves are fixed together we begin to unravel a fascinating story. The leaves are held together by sticky strands of silk, but the ants do not make silk themselves and it was some time before naturalists discovered the source of the silk. The ants get it, in fact, from their larvae, which they move back-wards and forwards like shuttles while fixing the leaves together. Before the leaves can be fixed, however, their edges must be brought together, and here the workers perform some clever acrobatics. Standing on their hind legs on the edge of one leaf, the ants reach out with their jaws and drag the neighbouring leaf down. Very often, however, the neighbouring leaf is out of reach of a single row of ants and the insects have to climb on each other's shoulders to reach it. Whole chains of ants may be needed to bridge the gap between two leaves, but they make it eventually and they gradually draw the leaves together. More workers then appear with the larvae and they move to and fro across the seam, dabbing the larvae down on each side to attach the sticky silk. The threads are so tightly packed that they seem to form a woven sheet between the leaves. Weaver ants live only in the warmer parts of the world and they bite fiercely when their nests are disturbed.

Trap-building insects

Apart from man and the spiders, very few animals make anything resembling a trap with which to catch their prey. Among the insects, the best known trap-builders are the larvae of certain caddis flies and ant-lions. The caddis larvae merely spin silken webs on the underwater vegetation and wait for the current to bring food along. The ant-lion traps are rather more elaborate, consisting of conical pits excavated in sandy soil. The ant-lion larva buries itself at the bottom and leaves just its massive jaws exposed. When an ant or some other small insect blunders into the pit the loose sand prevents it from scrambling out again and the ant-lion gets another meal. Neither the caddis larvae nor the ant-lion larvae do anything to lure their prey into the trap, but the New Zealand glow-worm definitely entices its victims with its glowing lamp. The insect is not related to the true glow-worms and it is the larvae of a little fly called *Arachnocampa luminosa*. It makes its home in caves and thousands of individuals normally live together. Each one makes a silken path on the roof of the cave and drops a number of vertical threads from it. The vertical threads are coated with blobs of gum and they hang down rather like little bead curtains. The glow-worm then retires to the path and glows steadily. The glow from the thousands of insects is reflected from the millions of sticky beads and tourists travel long distances to see the spectacle. But the insects do not light up for our benefit: they are interested in the small flies which are attracted by the lights. These flies are trapped by the sticky threads, and the glow-worms then haul up the threads for a feast.

The mating game

The majority of insects begin to advertise for mates as soon as they have completed their final moults, and most of them are very soon successful: it is very unusual to catch a female moth who has not already been fertilized and who will not oblige with a batch of eggs within a day or two. Such success is, of course, due to the efficient systems of communication that have evolved between the sexes. The messages may be in the form of scent, light, or sound —or even a combination of these—but whatever form they take, the receiver is able to pick out the signals of its own species from the host of other messages that surround it. Instinct then comes into play and the receiver reacts according to innate behaviour patterns which automatically draw it towards the source of the signals.

Nearly all butterflies and moths use scents in their courtship activities, and some of the most powerful attractants are products by female silkmoths of the family Saturnidae, including the European emperor moth. As soon as she has emerged from her cocoon and hardened her wings, the female protrudes a scent organ from her abdomen and allows the scent to be carried off on the breeze. The male moths have extremely elaborate antennae designed for picking up the scent, and they can actually detect just a few molecules of it in the air. As soon as a male's antennae are stimulated by the scent, he automatically turns upwind and thus begins his approach to the female. A lull in the breeze may mean that he loses the scent, and if this happens he will merely return to his random flying. There is no question of his feeling deprived and actually searching for the scent trail again, but he may well pick it up again and once more turn upwind. After perhaps several of these false starts, he will arrive at the female, but he often finds that he has to compete with numerous other suitors who have arrived to stake their claim. When once the female has accepted a male she stops emitting scent and the superfluous males gradually disperse. Experimental work has shown that the males can find their way to the females from several miles away, and they can do so even when the air is saturated with other scents. Only the scent of their own species causes them to fly upwind, and this reaction will always take place when the appropriate scent molecules are picked up, whether or not other scent molecules are hitting the antennae at the same time.

Light signals are used by several insects, but they are best developed in the glow-worms and fireflies, both of which are actually beetles. The light is produced by the oxidation of a substance called luciferin. The process takes place in special luminous organs and almost the entire energy output is in the form of light; there is virtually no heat generated. The insects can switch their lights on and off at will, possibly by controlling the oxygen supply to the luminous organs. Some species glow continuously while advertising for their mates, but others emit flashes at definite intervals. Light may be produced by either sex or by both sexes, but each individual is sensitive only to the wavelength or the flash pattern of its own species. The most amazing displays are those given by certain fireflies in south-east Asia. The males are drawn to each other by their flashing lights and they congregate in vast numbers on certain trees, usually along the river banks. Their proximity to each other somehow causes them all to flash in unison, so the whole tree seems to flash brightly about twice a second. The females are drawn to the 'firefly trees' and they have no trouble in finding mates. Because there are no seasons in this region and adult fireflies are maturing all the while, the new fireflies are drawn to established firefly trees and the same trees go on flashing night after night.

Sound signals are used by male cicadas, grasshoppers, and crickets. Cicadas 'sing' by vibrating tiny membranes on the sides of their bodies, while grass-hoppers rub their hind legs against their wings and crickets rub their front wings together. Each species has its characteristic sound, which is appreciated only by the females of the same species. Most of the songsters stop singing when they are disturbed, and their cryptic colours help them to escape detection, but the tree crickets have evolved an unusual method of deceiving

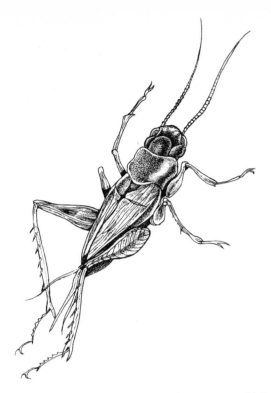

Crickets belong to the Order Orthoptera, which includes grasshoppers and locusts. Most of this Order have very long back legs, 'ears' (in crickets on their fore legs), and means of producing sounds. These sounds are made only by the male, as a means of attracting a mate, and in the case of true crickets are produced by grating a toothed file on the right forewing against the back edge of the left forewing. Females (like the one pictured here) possess long needle-like ovipositors

their enemies: they have become ventriloquists. Tree crickets live in southern Europe and many other parts of the world and the males put forth their sweet, bubbling song soon after nightfall. They sing from the trees and shrubs and it might be thought that tracking down such a clear song would be easy. It is not, as Fabre and many other naturalists found out as soon as they tried. When free from interference the male cricket raises his front wings so that they are perpendicular to the body and the sound comes over loud and clear. But walk towards the sound and it will appear to be coming from much further away. Carry on walking towards it and it will stop for a few seconds before suddenly starting up again behind you. Try again and you might find the sound suddenly switching to your right or your left. The cricket is not actually moving, but throwing its voice. The effect upon the searcher is, in Fabre's own words, complete comfusion. Presumably some of the cricket's natural enemies are similarly confused, but the voice-throwing does not affect the female because she would not venture forth while danger was in the offing. The voice-throwing mechanism is delightfully simple: as one approaches the singing insect, it lowers its wings and thereby reduces the volume. The effect is just as if the sound were now coming from further away. A slight twisting of the wings can beam the sound out to left or right.

Warning colours and mimicry

Many distasteful or otherwise unpleasant insects are boldly coloured or patterned and, far from hiding themselves from predators, they expose themselves freely on the vegetation. Familiar examples include the ladybirds, the wasps, and the gold-and-black ringed caterpillars of the cinnabar moth. The bold colours warn birds and other predators to leave the insects alone, but the predators have no inborn aversion to bold colours and they have to learn their significance. Young birds will peck at almost anything but, after trying a few boldly marked insects and finding them unpleasant, they become much more discriminating. The bold markings are obviously rapidly absorbed into the memory and they make a lasting impression on the birds. The efficiency of warning colouration has been shown in several striking experiments, including some with cinnabar catterpillars and mealworms. Young birds were given the cinnabar caterpillars first and they rejected them as soon as they had tried them. They were then given mealworms, a favourite with most birds, and they devoured the lot. The next stage was to give the birds mealworms which had been painted to look like cinnabar larvae: these were rejected without being tried. More normal mealworms were then offered and accepted, showing that the birds were still hungry. The birds obviously avoided the painted mealworms because of their appearance. Similar experiments have been carried out with lizards and with toads, all with similar results.

The commonest combination in warning colouration are black and yellow, black and red, and black and white, but it does not follow that all insects displaying such colours are nasty. Many harmless and edible insects have come to resemble unpleasant species and they use the resemblance, combined with the appropriate behaviour, to deceive their enemies. This form of deception is known as Batesian mimicry. The unpleasant species in each example is called the model, while the imitator is called the mimic. The phenomenon has evolved in exactly the same way as camouflage, slight resemblances to a model being improved over the generations until very strong likenesses have been achieved in some instances. Although some of these similarities really are remarkable, the resemblance does not have to be perfect to be effective: a slight resemblance that puts off just a few predators will help the insect species as a whole. But natural selection will, of course, always work to improve the similarity. The model is usually much more common than the mimic, for the obvious reason that predators must get to associate the shared pattern with an unpleasant experience and not with a nice one, but it is not essential that the model outnumbers the mimic. Experimental work has shown that an unpleasant experience has a much more marked and lasting effect on a predator and that it can erase the memory of several pleasant experiences.

False eyes and heads

Many insects, both young and old, employ false eyes to frighten their enemies. The 'eyes' are not usually prominent when the insects are at rest, and they may be completely concealed, but they are suddenly displayed when the insects are disturbed and their size is such that they appear to belong to a much larger animal. The effect on a predator, and on the inquisitive entomologist, is truly startling. Several moths indulge in this kind of behaviour, the most effective being those such as the eyed hawkmoths which have the eye spots on their hind wings and which expose them merely by raising the front wings a little. The same technique is employed by a large brown bush cricket from Brazil. Hanging upside down among the foliage, the insect is easily passed over, but disturb it and you are immediately confronted with what seems to be a glowering owl. The eye spots, the mottled brown wings, and the body are all involved in this remarkable piece of bluff. The larvae of several kinds of hawkmoths imitate snakes when they are disturbed and they gain protection in that way. They withdraw their heads into their bodies, thus enlarging the thoracic region and expanding the eye spots there. The illusion is completed by a sinister waving of the body from side to side.

The wings of many butterflies, notably the 'browns' (family Satyridae), bear small eye spots near the margins. These spots appear to act as decoys, drawing the attention of predators away from the more vulnerable parts of the body. Some of these butterflies actually seem to use their eye spots rather like the soldier who pushes his helmet up from his hide-out to see if anyone shoots at it. When the butterfly settles on the ground it often leaves a prominent eye spot exposed for a few seconds as if to invite attack. If none is forthcoming, the butterfly apparently concludes that its resting place is safe, hides the eye spot away by folding its wings down, and merges in with the surroundings.

Some insects use the decoy system in an even more elaborate form, fooling their attackers by having prominent false heads at the hind end of the body. The wings bear small eye-like markings and often carry outgrowths like antennae. Predators must be completely misled by such adornments. Entomologists, too, have been disappointed, approaching their quarry stealthily only to find it flying off in completely the opposite direction.

Jamming the bat's radar

Visual camouflage and bluff are fine when directed against diurnal predators, but they offer no defence against the night-flying bats. These animals snap up all kinds of flying insects, which they find by echo-location. Some moths, however, have evolved anti-bat defences. These moths have hearing organs and they can detect the high-pitched sounds emitted by the bats. If the sound is very intense, meaning that a bat is close at hand, the moths generally drop to the ground out of harm's way, but if the bat is further off the moths merely turn away from the direction of the sound. Some of the moths change their wing-beat frequency and this may well upset the bat's radar, but others, notably the tiger moths, actually emit ultra-sonic pulses of their own when they pick up the bat's signals. Further research is necessary to find out whether the moths' signals actually jam the bat's radar or whether they merely act as a warning to the bat. Tiger moths certainly possess poisonous glands and spines and it may be that the pulses which they emit are the auditory equivalents of their warning colouration. The moths are not concerned with the technical details, however: the important thing for them is that the sounds protect them from the bats. This certainly happens.

Strange partnerships

We have already seen how certain ants form mutually beneficial associations with aphids, but the aphid is actually only one of many kinds of creatures that take up residence in ant nests. Well over 2000 species, most of them beetles, have been discovered living with ants in various parts of the world. These are not all welcome guests, of course, and some of them are positively harmful to the ant colonies, but one which definitely is welcome, and which is actually brought home to tea by the ants, is the caterpillar of the large blue butterfly. The caterpillars of many blue butterflies (family Lycaenidae) secrete honeydew and are attended by ants, but the large blue caterpillar is obviously something rather special as far as some of the European red ants are concerned.

The large blue butterfly is widely distributed in Europe, although extremely rare in the British Isles, and its caterpillar feeds on wild thyme in its early stages. After the third moult, however, it loses interest in its food and wanders about on the ground. A gland on the abdomen has begun to exude honeydew by this time and the caterpillar attracts numerous ants. Eventually a small red ant manages to drag the caterpillar back to its nest, where the guest receives first class attention. It is installed in a comfortable chamber and its taste for meat is satisfied by a liberal supply of the ants' own larvae. The ants ask nothing in return except that the caterpillar goes on secreting the honeydew which they love so much and which they continually lick from its body. The caterpillar pupates in the nest and the ants later allow the emerging butterfly to pass unhindered out into the air. A few blue butterflies in other parts of the world have similar life histories.

Some ants actually go into partnership with plants. Several kinds of acacia trees, for example, possess swollen thorns which are hollowed out and occupied by small ant colonies. The ants feed on sugary and oily secretions from the stems, and they repay the plants by keeping them free from insect pests. Some of these ants sting or bite so viciously that even man thinks twice about wielding his axe on the trees.

A great many insects feed on the nectar and pollen of flowers, and most flowers are cleverly designed so that the foraging insects will pollinate them at the same time. No special action is necessary on the part of the insect as a rule, but there are instances in which the insect goes out of its way to ensure that the flowers are pollinated. Needless to say, such actions are not entirely altruistic, for the survival of the insect species is at stake as well. The spiky yucca plants of Mexico and neighbouring regions maintain a very close relationship with some small moths of the genus *Pronuba*, and neither plants nor insects can reproduce without the other. The proboscis of the female moth has become modified into a pair of curved tentacles with which she industriously collects pollen from the yucca flowers. This pollen is slightly sticky and the moth gradually works it into a ball two or three times the size of her own head. Balancing this load between her front legs and her head, she flies off to another flower and, using her long ovipositor, she lays some eggs in the ovary of the flower. The grubs that hatch from the eggs are destined to feed on some of the yucca's seeds, but the seeds will not grow unless the flower is pollinated: the yucca moth makes sure that it is. She moves to the stigma of the flower and rubs her ball of pollen over it. Sufficient pollen sticks to the stigma to fertilize numerous ovules and produce numerous seeds. The yucca moth caterpillars will eat some of the seeds, but there will be plenty left to scatter and produce new yucca plants. The behaviour of the female yucca moth is perhaps the nearest we find to a deliberate, reasoned action in the insect world, for she certainly appears to be thinking ahead. In reality, of course, she is merely following an instinctive pattern, a pattern which has evolved because moths possessing this sort of behaviour have been the most successful at rearing offspring.

The beautiful flower mantis need do no more than stay still to attract its prey. When an approaching insect comes within range it is speared on spines growing from the mantis's legs. The eye spots on the wings scare off would-be predators. The female also has the delightful habit of eating her man after mating

Left The huge nests of the highly destructive termites are also efficient and complicated homes, with air-conditioning, central heating and organically broken down food. A nest will contain workers, winged males and females, nurses and soldiers (*below*). The queen, seen here tended by workers (*right*) who feed her, groom her and carry away the eggs, is simply an egg-producing unit. At times she is capable of producing 30 eggs a minute, and may live for 50 years

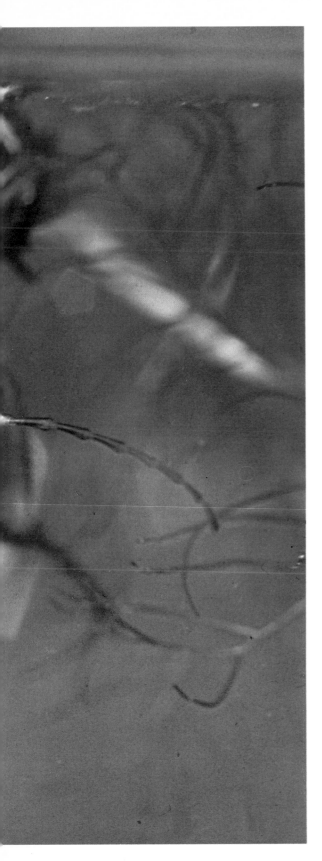

Left The giant diving beetle (*Dytiscus*) manages to breathe below water by trapping air under the wing covers of its streamlined body. Its 2in. larvae have huge jaws, with which they will attack and devour anything available—especially tadpoles and small fish

Above This ichneumon fly (*Rhyssa persuasoria*) is a parasite on the giant wood wasp and will drill through the wood of pine trees to reach its prey. Its ovipositor, clearly seen here, is more than twice as long as its body length

Below Some insects have to tolerate unwelcome passengers. Here a dipteran fly is carrying a pseudoscorpion, which is not an insect but is related to spiders and scorpions. Tiny (less than $\frac{1}{4}$in.), pseudoscorpions feed on springtails and other small invertebrates, which they kill and tear apart with their venomous pincers

The eggs (*above*) of this *Culex* mosquito are laid in rafts on the water surface. When they hatch, the larvae (*right*) hang head down just below the surface, feeding on tiny particles. They breathe air through a narrow tube to the water surface. The pupae (*right centre*) hang like commas, also below the water surface. Male mosquitoes do not suck blood, and (*below*) a male is seen just emerging from its pupa. After mating it will die, but a female will live for several weeks, taking several blood meals, which are essential for the several batches of eggs she will lay

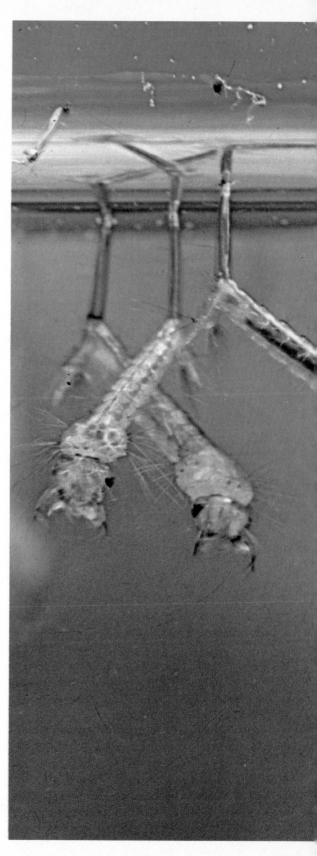

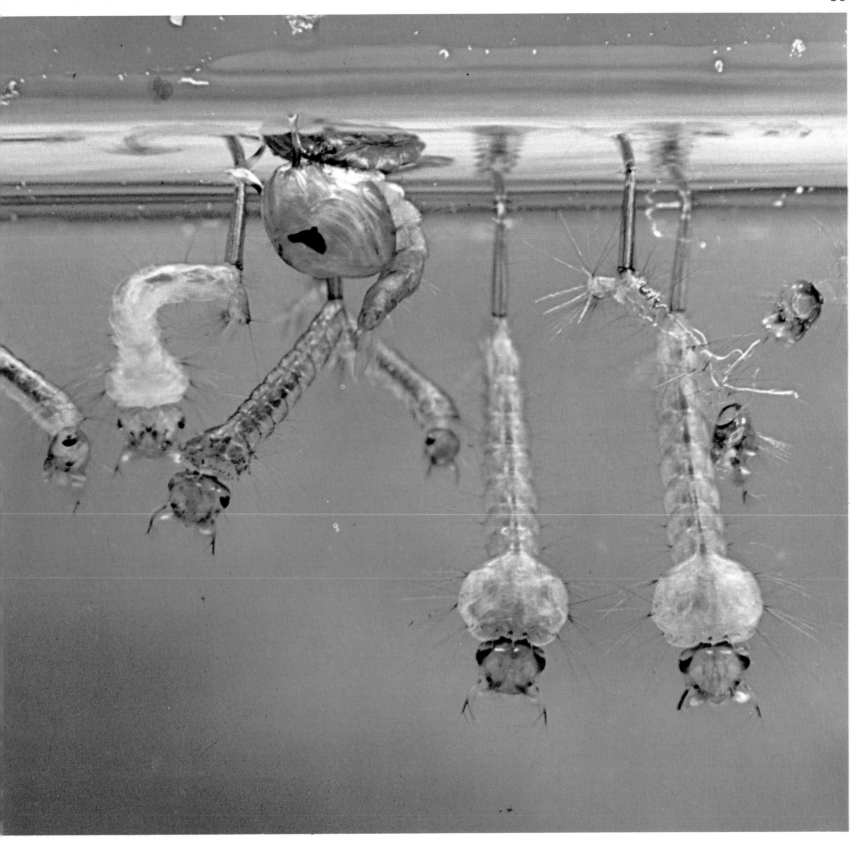

Evolution

by Dr Roger Hamilton

Dragonflies have no pupal stage. The nymphs, which resemble wingless adults, continue to develop for from one to five years, according to species. When ready, they climb out of the water, shed their last nymphal skin, expand their wings and fly as adults, leaving behind their nymphal skeleton

Mammals

The mammals originated from reptile ancestors about 200 million years ago. The earliest known remains of mammals are from Lesotho in Southern Africa and Glamorgan in Wales. These remains show that the first mammals looked very like living shrews. They fed on insects and were probably active only at night. For the first 120 million years of their history the mammals lived in a world that was dominated by the dinosaurs and throughout this time all the mammals remained small and probably competed with lizards for their food. During this period however the mammals split into two major groups: the marsupials or pouched mammals, which survive in Australia and South America; and the placental mammals, which survive as most of the other

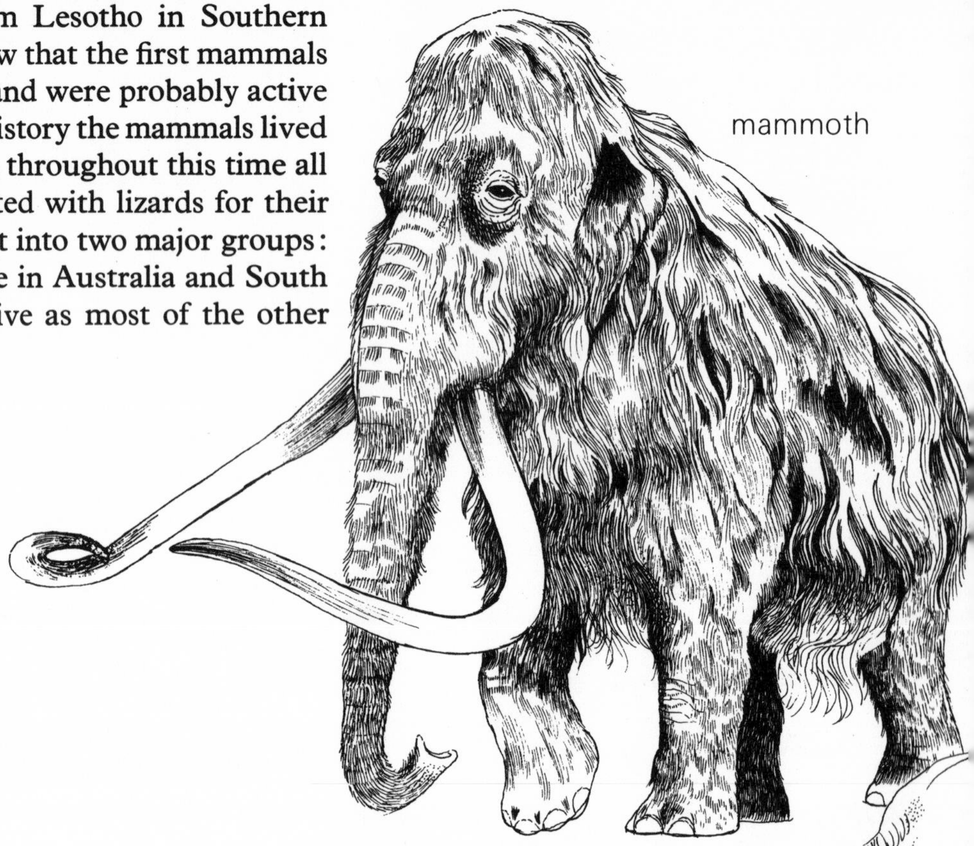

mammoth

Irish elk

Uintatherium

mammals in the world.

The dinosaurs became extinct about 70 million years ago and this left the way open for the mammals to take over as the dominant land animals. During the following 20 million years they expanded very rapidly, producing a group of early flesh-eating animals that included mammals similar to living cats and dogs. Several groups of large plant-eating mammals also evolved at this time and one of these included *Uintatherium* which was about the size of a rhinoceros and had sabre-like tusks, large horns and the smallest brain, relative to its body weight, of any known mammal.

Between 35 million and 20 million years ago the ancestors of many living mammals were evolving. Early elephants, hyraxes, whales and sea cows lived in North Africa 30 million years ago, while in North America early horses, rhinoceroses and cats were evolving. The first horse—eohippus (*Hyracotherium*)—is known from North America and Europe in rocks that are 50 million years old. Eohippus was about the size of a terrier dog. It had three toes on each of its back feet and four on each front foot. This small horse fed on soft plant material and lived in the dense tropical forests that grew in Europe and North America at the time. The descendants of eohippus lived on the plains of North America and much of the evolution of the horses occurred in the New World with waves of progressively more advanced horses invading Asia during the 50 million year period that is spanned by the evolution of the horses.

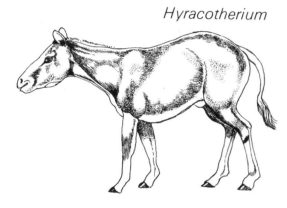

Hyracotherium

Ten million years ago a huge stretch of plains and grasslands stretched across the Old World from western Europe and north Africa as far as China. Some of the mammals living on these plains were very similar to those of Africa today. There were gazelles, antelopes, rhinoceroses, big cats, hyenas and giraffes. Many mammals now extinct also lived on these plains including three-toed horses of the *Hipparion* group and *Deinotherium* which is a relative of the elephants but had large tusks only in its lower jaws. There was also a huge relative of the horses and rhinos that had big claws on its feet. These claws were used to dig for roots and tubers which formed the animal's food.

About one million years ago the Ice Ages began and this had a dramatic effect on the mammals of the world. At the same time man was becoming an important hunter and this also affected the mammal fauna of the world. There is no known reason for the Ice Ages, but during the last million years the Arctic Ice Gap has expanded southwards at least three times, reaching almost as far south as London at its maximum extent, only to retreat northwards again during warmer periods. In cold periods mammoths, woolly rhinoceroses, polar bears, reindeer and wolves lived in Britain and southern Europe, while in warm periods elephants, giant deer, hyenas, cave lions, sabre-toothed cats, hippopotamus, rhinoceroses and horses were among the exotic mammals that lived in Europe. In later Ice Age times an unexplained wave of extinctions occurred. This affected mainly the larger mammals and the survivors of these extinctions form the living mammalian fauna of the world.

Deinotherium

Birds

Birds originated about 150 million years ago from a group of reptiles that is very closely related to the dinosaurs and crocodiles. The earliest known fossil bird is *Archaeopteryx*, which is in many features transitional between the reptiles and advanced birds. The bones of *Archaeopteryx* are solid as in the reptiles, whereas in birds the bones are hollow and contain air sacs that make the skeleton lighter. This is an adaptation for flight. The skull of *Archaeopteryx* has a short face and its jaws carry rows of small teeth, whereas all advanced birds have beaks. In general appearance the skeleton of *Archaeopteryx* is very like that of a small reptile and this similarity is made more striking by the presence of a long bony tail, a feature that was lost very early in the history of

Ichthyornis

Archaeopteryx

Phororhacos

Aepyornis

Hesperornis

the birds. Even the wings of *Archaeopteryx* are very small and their bones are like those of the front legs of a bipedal reptile. *Archaeopteryx* was however definitely a bird, as the impressions of many well developed feathers are preserved round the fossilized skeleton, and feathers only occur in birds.

There is a gap of about 60 million years in our knowledge between *Archaeopteryx* and the next occurring fossil birds. During this time the birds had expanded greatly and over twenty species of birds occur in rocks between 80 and 90 million years old. These include a small sea bird, *Ichthyornis*, which probably had a way of life similar to the living terns. Another well known sea bird of this time was *Hesperornis*, which was about 6ft long and was a very specialized sea bird. It was flightless and had entirely lost its wings. *Hesperornis* swam using its large back feet. The head of *Hesperornis* carries a long face with small teeth along the upper and lower jaws. These teeth would have allowed *Hesperornis* to hold struggling fishes.

Most of the groups of birds now living had evolved by 70 million years ago and at about this time the birds radiated very rapidly. The most important division of living birds splits the group into those that can fly and the flightless forms such as the ostrich and emu. Flying birds of 50 million years ago include the herons, gulls and pelicans. These birds lived near water and were therefore frequently fossilized, but there is no doubt that a wide radiation of birds was also occurring in drier habitats. Fossils of flightless birds are very abundant in Madagascar and flightless birds seem to have survived well on this island. Fossil birds from Madagascar include *Aepyornis*, the elephant bird, which reached a height of over 10ft. Eggs of *Aepyornis* are still found in the sands on the shores of Madagascar and they are the largest known eggs with a volume of over two gallons. Flightless birds were also very successful in South America where they had a way of life similar to large flesh-eating mammals. *Phororhacos* lived in South America about 15 million years ago. It was about 10ft tall and had a skull that was 16in long with a huge deep, hooked beak suitable for killing and for tearing flesh.

The fossil record of the birds shows their radiation throughout the past 50 million years to their present diversity and worldwide distribution. There has however been some decline in the last million years as a result of the Ice Ages and hunting by man.

Fishes

The fishes are the most important aquatic vertebrates. Their evolutionary history is well known as they have a good fossil record that stretches back into rocks that are over 420 million years old.

The earliest fishes had not yet evolved bony jaws, and the living lamprey is a soft bodied representative of this group of jawless fishes. Most fossil members of the group had heavy bony armour that covered and protected their bodies but probably also meant that they were poor swimmers. *Cephalaspis* had very heavy armour and is thought to have lived by extracting food from the muddy beds of streams and rivers. Later armoured fishes such as *Bothriolepis* had well developed jaws and were probably active swimmers.

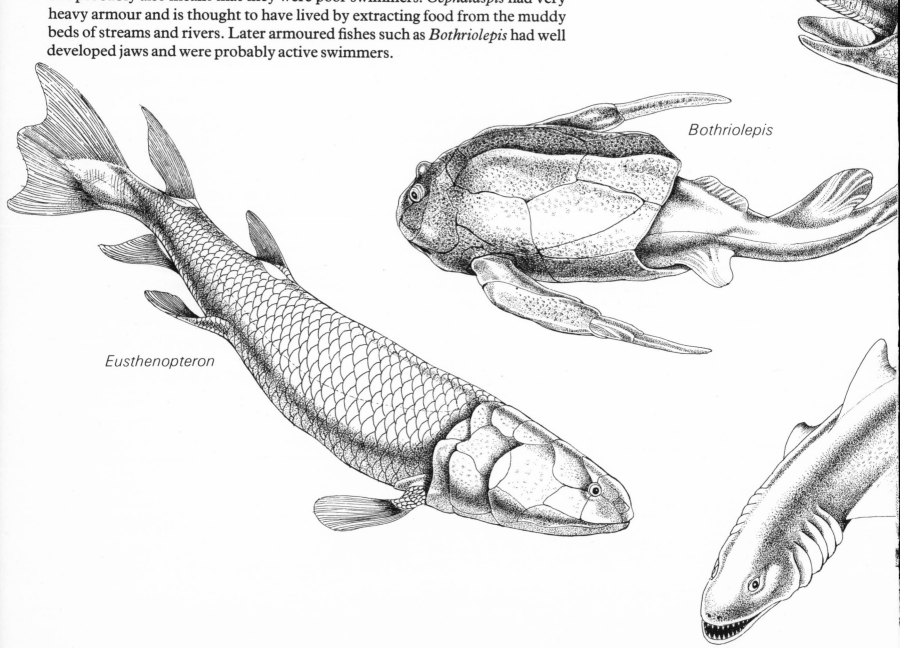

Bothriolepis

Eusthenopteron

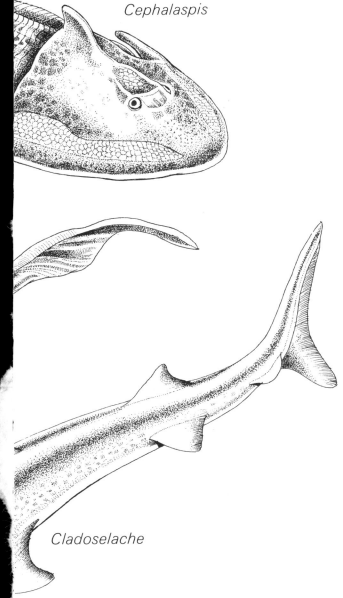

Cephalaspis

Cladoselache

The earliest sharks appeared about 370 million years ago. Sharks and rays have skeletons made of cartilage, which is a gristle-like substance that does not fossilize very well. Sharks' teeth and fin spines are, however, abundant as fossils. The sharks were the last important group of fishes to evolve, and *Cladoselache* is an early member of the group. Throughout their later history the sharks radiated, improving their teeth for killing and slicing flesh or developing low flat teeth that are suitable for crushing shellfish.

The advanced bony fishes are by far the most diverse vertebrates and with over 20,000 species they outnumber all other vertebrate groups combined. These bony fishes probably first appeared in freshwater about 400 million years ago. They were successful almost immediately and quickly expanded into the sea. One group of bony fishes has ray-like fins. These 'ray-finned' fishes are the most common living members of the group and they include the cod, mackerel, herring, trout, salmon, stickleback and a host of familiar marine and freshwater fishes. Fossils of ray-finned fishes indicate that the group was enjoying moderate success as much as 200 million years ago, but since then it has expanded enormously so that the ray-finned fishes are now at the peak of their diversity and abundance.

Although less important in fish history, the 'lobe-finned' fishes are particularly interesting as they include the ancestors of all land vertebrates. *Eusthenopteron* is one of the best known fossil lobe-finned fishes. This animal lived about 360 million years ago. The fins of *Eusthenopteron* consist of a fleshy base or 'lobe' which is muscular. The fin projects from this and can be moved backwards and forwards by its base. In *Eusthenopteron* these fins could have been used to assist the sinuous movements of the body thus allowing the fish to 'walk' on land. The living coelocanth is a lobe-finned fish that was thought to have become extinct about 100 million years ago. It was, however, discovered in 1938 living in deep water off the coast of Africa.

The modern fish fauna of the world is dominated by the ray-finned bony fishes, with sharks and rays also abundant. Other groups such as the jawless lampreys, the strange lung fishes and the coelocanth remind us that the fishes had a complex history.

Amphibians

Living amphibians include frogs, toads, newts and salamanders. They are a relatively unimportant part of the world's fauna today, but they were the most important land animals for a period of 80 million years before the reptiles became important. The amphibians originated from the lobe-finned fishes and *Eusthenopteron* may have been closely related to their ancestors. One of the earliest amphibians was *Ichthyostega* which was up to 3ft long and looked like a large newt. *Ichthyostega* is found in rocks that are 360 million years old that occur in Greenland. It had a large head and along its tail there is a fish-like fin. The limbs of *Ichthyostega* were, however, well developed and there is no doubt that it could have walked on land, although it may have spent most of its life in water.

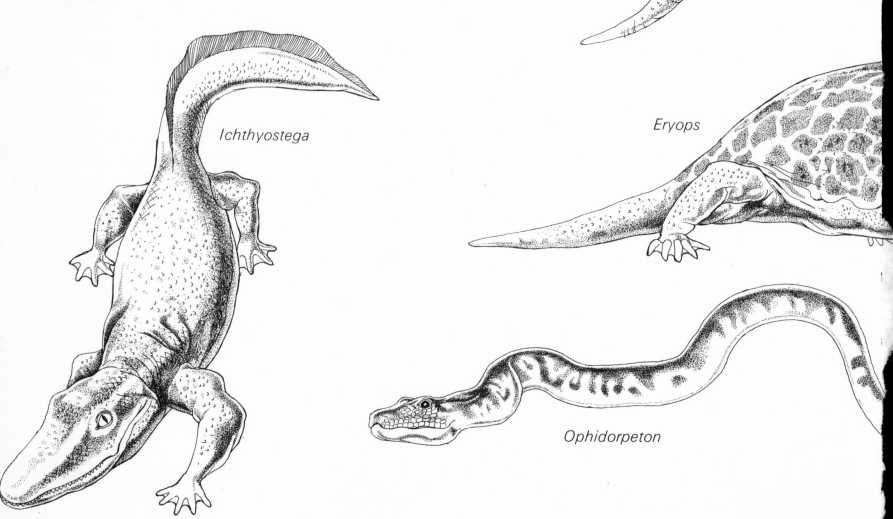

Seymouria

Ichthyostega

Eryops

Ophidorpeton

Later amphibians were very successful but most of them remained amphibious, while some lived entirely in water. The amphibians were the largest vertebrates living in the coal forests, where they probably fed on fishes, insects and each other. Among the larger coal forest amphibians was *Megalocephalus* which was over 3ft long. There were also many smaller amphibians including snake-like forms and the huge almost limbless *Pteroplax*, which had a rounded body up to 16ft long. *Pteroplax* probably lived in water and fed on fishes.

The amphibians gave rise to the reptiles about 300 million years ago and the flesh-eating *Seymouria* may have been closely related to the reptile ancestor. *Seymouria* was less than 3ft long. It had a relatively small head and well developed legs that would have allowed it to walk easily on land. The amphibians continued to be successful for the early part of reptile history and may have competed for food with the early reptiles. *Ophidorpeton* was an advanced snake-like amphibian with a slender body up to 3 ft long while *Eryops* was almost 6ft long and must have been a clumsy land animal.

The amphibians tended to decline as the reptiles became important, but the frogs, newts, salamanders and legless amphibians known as caecilians survive. Frogs and toads have very unusual bodies that are specialized for jumping, with long back legs, no tail and a very short back-bone. Most modern groups of frogs were already in existence 100 million years ago, but fossil frogs as much as 180 million years old differ little from living forms. The earliest fossil frogs come from Madagascar and are about 200 million years old, but they provide no clues to the ancestry of the group. The newts and salamanders have bodies similar to those of many early amphibians, which presumably reflects a similar way of life. The oldest salamander is 130 million years old but again it is similar to living forms and gives no clues to the early relatives of this group. The caecilians are burrowing amphibians that look like large earthworms. They have no fossil record and their history is therefore unknown.

Reptiles

Living reptiles include the snakes, lizards and crocodiles. They are now a relatively unimportant part of the world's fauna. The reptiles were, however, dominant on land and in the sea for the huge time span of over 200 million years before the rise of the mammals. Reptiles originated from amphibian ancestors about 280 million years ago at the time that the coal forests were growing and forming one of our most important fossil fuels.

The key feature that led to the success of the reptiles was their development of a shelled egg that could be laid on land, and inside which the young reptile developed until it could live on land. This freed the reptiles from close dependence on water and allowed them to expand into the drier parts of the

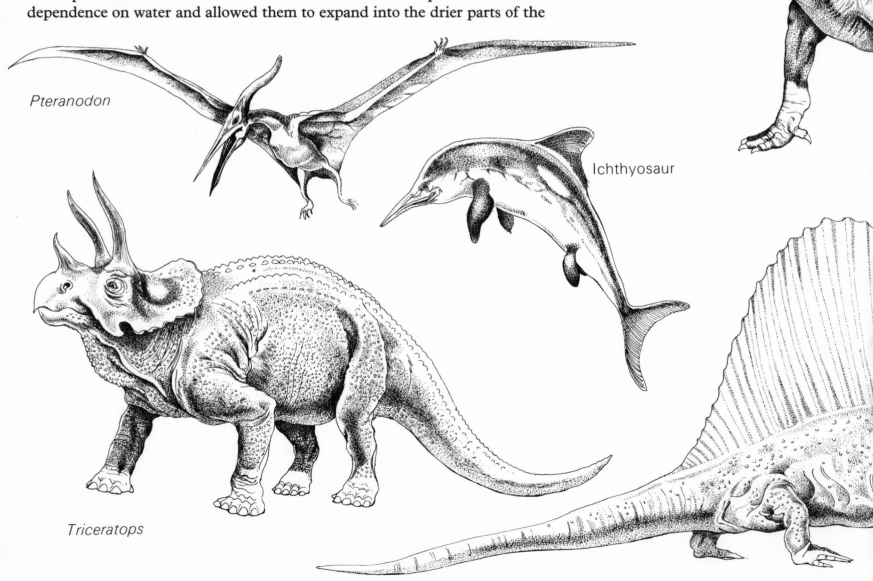

Pteranodon

Ichthyosaur

Triceratops

Tyrannosaurus

Dimetrodon

Earth. Early reptiles had bodies like huge lizards. *Dimetrodon* was a large flesh-eater that had a sail-like growth on its back. This sail may have helped *Dimetrodon* to absorb or radiate heat and may have been an early attempt to regulate body temperature. *Dimetrodon* belonged to a group known as the mammal-like reptiles. These animals were dominant on land for about 60 million years and they gave rise to the mammals about 200 million years ago. At about this time the mammal-like reptiles declined rapidly and were replaced by the dinosaurs and their close relatives.

The ancestors of the dinosaurs ran only on their back legs, i.e. they were bipedal, and many later dinosaurs such as *Tyrannosaurus* and *Iguanodon* retained this way of running. All flesh-eating dinosaurs were bipedal; they ranged from the size of a hen to huge beasts such as *Allosaurus* and *Tyrannosaurus*, which was the largest flesh-eating animal that has ever lived on land. A small bipedal dinosaur known as *Ornithomimus* was about the size of an ostrich. It had a long neck, long legs and a long tail. Its head was very small and *Ornithomimus* had a beak rather than teeth. It has been suggested that *Ornithomimus* fed on the eggs of other dinosaurs and it must have been a very fast runner to survive this dangerous way of life. Huge plant-eating dinosaurs walked on all four feet. *Diplodocus* is the longest land animal that has ever lived, while *Brachiosaurus* weighed about 50 tons and is the heaviest land animal known. *Stegosaurus*, *Ankylosaurus* and *Triceratops* were all heavily armoured dinosaurs that may well have been able to defend themselves against their flesh-eating cousins.

While the dinosaurs were ruling the land, other reptiles were important in the seas. Turtles are first known from rocks that are 200 million years old and inside their armour they have been little affected by the passage of time. Ichthyosaurs were very fish-like in appearance with dorsal fins and high flattened tails that gave their power for swimming. Ichthyosaurs could not walk on land and they gave birth to live young in the water. This is indicated by fossils which show ichthyosaurs in the process of giving birth. Plesiosaurs were like huge seals but they had long necks and small heads. They fed on fish and swam using their paddle-like limbs.

In the air the pterosaurs such as *Pteranodon* and *Pterodactylus* enjoyed a long period of success. These strange reptiles may have had a coating of hair. They had large wings that consisted of flaps of skin supported by greatly lengthened finger bones. One recently discovered fossil pterosaur had a wing span of over 50ft. Pterosaurs were probably not very good fliers but at the time there was nothing better. They may have fed on fish and it has been suggested that they lived along cliffs where the wind would have helped them to take off for flying.

About 70 million years ago a huge wave of extinctions occurred and most marine reptiles, all the dinosaurs and the pterosaurs became extinct. The turtles, lizards, snakes and crocodiles survived into a world where the mammals were expanding.

Insects

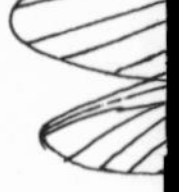

Most insects live on land and all adult insects breathe air, which suggests that the group evolved from a land-dwelling ancestor. The earliest known fossil insects are from rocks that are about 370 million years old and they were discovered at Rhynie near Aberdeen in Scotland. These very early insects are similar to the living springtails and they do not have wings. Another group of wingless insects includes the living silverfish, and relatives of this small insect first occur in rocks about 60 million years old, though fossils 300 million years old may possibly belong with this group.

Winged insects are known from rocks that are 300 million years old and they were very common in the coal forests. The most primitive of these insects

Metoedischia

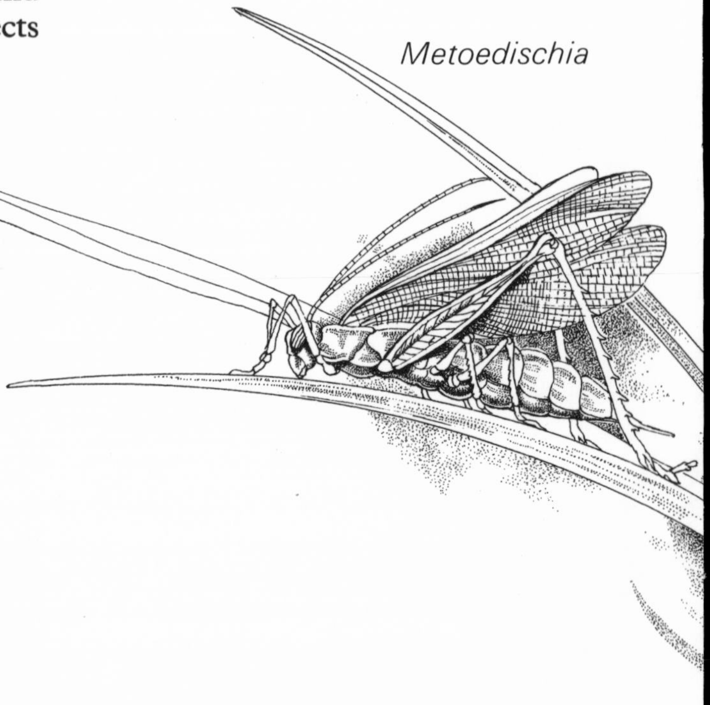

Meganeura

insect in amber

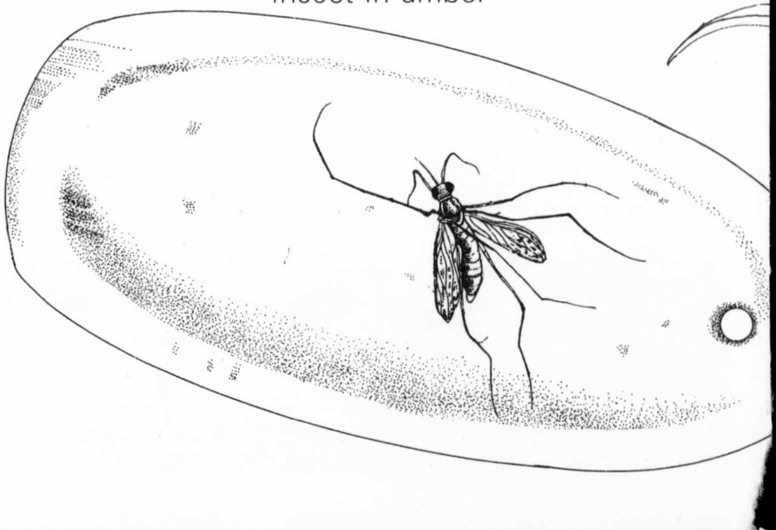

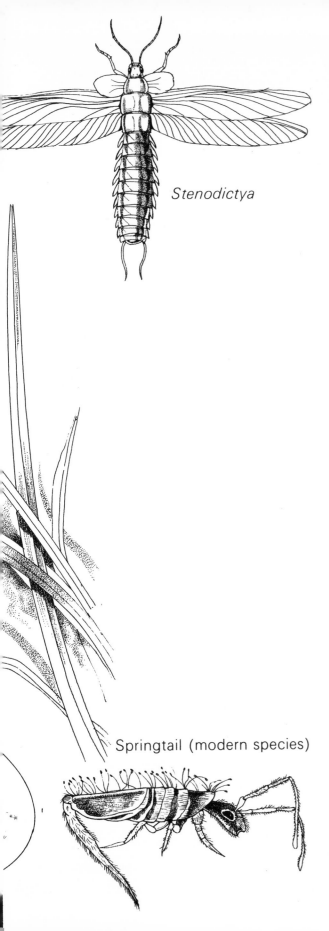

Stenodictya

Springtail (modern species)

have two nearly identical pairs of wings which project outwards from the sides of the body. The wings are permanently spread and cannot be folded backwards. *Stenodictya* is one of the most primitive of these early winged insects. It has small flaps at the front of its thorax which represent a rudimentary third pair of wings. The veins of the wings are very numerous and spread over all parts of each wing, whereas in more advanced insects there are fewer veins and they have a more restricted distribution over the wing. *Stenodictya* also had a frill of small projections around its abdomen; these do not occur in more advanced insects.

True dragonflies are known from rocks over 200 million years old and a group closely related to their ancestors occurs in rocks that are about 300 million years old. These include many insects that were like dragonflies and one of the best known in *Meganeura*, which is the largest flying insect that has ever existed. It looked very like a modern dragonfly but had a wing span of from 60 to 75cms (2ft-2ft 6in.).

Cockroaches first occur in rocks that are 300 million years old, and are common in many rocks younger than 250 million years. They are more advanced than the dragonflies because they can rotate their wings backwards so that they partly overlap, with the front pair over the back pair. Fossil cockroaches that are found preserved in coal deposits are essentially the same as those living today. The earliest crickets also occur as fossils in coal deposits but the closely related grasshoppers do not appear until much later. Crickets such as *Metoedischia* from rocks 250 million years old are very similar to living members of this group.

The beetles are a very important and diverse group of living insects with over a quarter of a million species. The oldest known fossil beetles occur in rocks that are about 250 million years old, but for such a large group the beetles are surprisingly poorly represented as fossils. Many complete beetles are known as fossils, but usually only their hard wing cases are preserved. The evolution of these insects is not clearly understood.

The best known fossil insects are those preserved in amber, which is itself the fossilized resin of trees. Amber from the Baltic area of Europe is about 35 million years old. Many complete insects including lacewings, flies and ants have been found embedded in this amber with their complete bodies and even colours preserved.

Although the main insect groups are known to have existed over 250 million years ago, the group has expanded steadily in diversity. The appearance of flowering plants gave a great stimulus to the evolution of the insects and between 200 million and 70 million years ago they radiated very rapidly in response to the rapid evolution and radiation of the plants which provide food and homes for many insects.

Index

Figures in italics refer to illustration pages

adder 91; *102*
adder, horned puff 103
adder, puff 103
Adélie penguin 74, 75
Aepyornis 165; *164*
Aesculapian snake 193
African elephant 13, 24–25; *14, 33, 34, 35*
albatross 67–8, 76
albatross, black-browed 74
alligator 97–8; *106–7*
alligator snapping turtle 96
Allosaurus 171
alpaca 28
alpine salamander 89
amberjack *125*
Ammodytes 86
Amphibian 89–91, 170–1
anchovy 113, 119
Andean condor *66*
anemone fish 115
anemone fish, yellow-faced *130*
anglerfish 114, 120, 128, 150; *127*
anhinga *87*
Ankylosaurus 171
ant 8, 141–2, 144–7, 173
anteater 48–9
anteater, giant 14, 49; *49*
anteater, scaly *48, 60*
anteater, spiny *49*
antelope 13, 15, 16, 31, 163
antelope, blue 9
antelope jack rabbit 45
Antilles frog 89
ant-lion 147
ant, weaver 146
anura 90
ape 17, 19
aphid 8, 139, 145
Arabian camel 28–30; *28*
Archaeopteryx 164; *164*
arthropods 137
Asian elephant 24–5
Asian green whip snake 104
Atlantic salmon 117
auk 75
auk, great 9
axolotl 93; *108*

baboon *59*
Bactrian camel 28–30; *28*
badger 53
badger, honey 52–3
baleen whale 13, 25
basking shark 114
bat 13, 44, 51–2, 151; (sonar) *52, 62*
bat, fish-eating 51; *63*
bat, free-tailed 52
bat, horse-shoe 51–2
bat, spear-nosed 51–2
bat, vampire 13
bear 20–22
bear, grizzly 21
bear, polar 20–1, 163; (distribution) *20, 21, 36*

beaver 41–3; *42*
bee 139, 140, 141–4, 146; *143*
bee-eater 28
bee-hummingbird 66
beetle 138, 140, 148, 173
beetle, giant diving *156*
Bengal monitor 102
billfish 114
bird 13, 65–8, 164–5
bird of paradise 70
bison 15
blackbird 71
black-browed albatross 74
blackcap 10
black-footed ferret 50
black grouse 70
black rat 56
black rhinoceros 27–8
blenny 123
blesbok 16
blue goose *84*
blue-footed booby 76
blue shark 116
blue streak 125, 126
blue whale 13, 26
booby 76
booby, blue-footed 76
booby, masked 76
booby, red-footed 76
Bothriolepis 166
bovid 31
bowl-backed tree toad 89, 95
Brachiosaurus 171
bristlemouth fish 114
brotulid 114, 126
brown lemming 46–7; *47*
budgerigar 73
buffalo 23
bushbuck 16
bush-cricket 7
bushmeat 16
butcherbird 78
butterfly 137–8, 140, 147–150, 151–2
butterfly fish 115, 119, 125
butterfly fish, four-eyed 119

caddis-fly 147
caecilian 89, 90, 171
camel 28–30, 31, 45; *28*
Canadian timberwolf 54
Cape pangolin *48, 60*
capelin 116
capybara 55
cascaval 103
cat 13, 20, 22–3, 53, 100, 162, 163
caterpillar 138, 149, 151
catfish 118
catfish, African 120
cattle 13, 15, 16, 31
cave-fish, Mexican 118
cave-olm 93
cayman 97, 98
centipede 137
Cephalaspis 166; 167
cervid 31
cetacean 25
chameleon 92, 100–1; *109*
chameleon, flap-necked 100
chameleon, Hoehnel's 100
chameleon, Jackson's 101
chameleon, mountain 101
chaffinch 71, 73, 79
cheetah 15; *59*
chevrotain 31
chevrotain, lesser Malay 31

chicken, prairie 70
chicken snake 103
chihuahua 54
chimaera 113
chimpanzee 17, 19, 20; *18*
chrysalis 140
cicada 148
Cladoselache 167
cleaner fish 123–6
cleaner wrasse 123
clingfish 113
cobra 92
cockroach 173
cod 114, 116
coelenterates 115, 124
common skink 91
common toad 90
condor, Andean *66*
cormorant 75
coyote 54
crab 98, 137
cricket 148, 150; *149*
crocodile 91, 92, 97–8, 101, 164, 168, 169
crocodile, dwarf or broad-fronted 98
crocodile, estuarian 98
cuckoo 73, 78, 146
Culex 158–9
curlew, Eskimo 80

dab 120
damselfish 125
darter 75
Darwin's finches 66
deer 13, 31, 43, 163
deer, mouse 31
deer, pig 101
deer, red *15*
deer, roe 44
deer, white-tailed (Virginian) *61*
Deinotherium 163; 163
desert fox 45, 53
desert monitor 102
desert pygmy monitor 102
desert rat 45
Dimetrodon 171; *171*
dingo 54
dinosaur 25, 65, 162–3, 163, 169
Diplodocus 171
dodo 9
dog 13, 53–5, 100, 162
dog, prairie 50
dolphin 25
donkey 45
dragon, Komodo 101; *105*
dragonfish 120, 128
dragonfly 139, 173; *160*
dromedary 28, 29; *28*
duck 67
duck, Labrador 9
duckbilled platypus 43; *43*
dugong 13
dung beetle 45
Dytiscus 156

eagle 67; *81*
eagle owl 47, 50
echidna 14, 49; *49*
edible frog 90
eel, electric 122
Egyptian vulture 77; *77*
eland 16
electric eel 122
elephant 14, 24–5, 163
elephant, African 13, 24–5; *14, 33, 34, 35*

elephant, Asian 24–5
elk, Irish 162
Elliot's storm petrel 76
emerald monitor 101
emperor penguin 71; *71*
Eohippus 163
Eryops 168; *169*
Eskimo curlew 80
Etruscan shrew 13
European tree-frog 90
Eusthenopteron 167, 168; *166*
eyed hawkmoth 150

falcon 78
fennec fox 45, 46, 53; *46*
ferret 44
ferret, black-footed 50
finch 67
finch, Darwin's 66
finch, warbler 66
finch, woodpecker 66
firefly 148
fish 113–123, 166–7
flamingo 67
flower mantis *153*
fly 138–9, 140, 147, 173; *140*
flying fish *122*
'flying snake' 104
fox 20, 43, 47, 53–5
fox, desert 45
fox, fennec 45, 46, 53; *46*
fox, red 53, 54; *14*
free-tailed bat 52
freshwater leaf-fish 124–5
frog 89–90, 95, 98, 99, 170, 171; (life cycle) *94*
frog, Antilles 89
frog, edible 90
frog, European tree 90
frog, grey tree 90
frog, marsh 90
frog, marsupial 89, 90, 95
frog, poison arrow *88*

Gaboon viper 92, 102–3
Galapagos giant tortoise 96
Galapagos storm petrel 76
gannet 74, 76
gavial 98; *97*
gazelle 15, 23, 163
gazelle, Thomson's 23
gecko 91, 104; *101*
Georgia blind salamander 94
giant anteater 14, 49
gaint panda 21–2
giant rat 16
giant salamander 90
giant squid 25, 26
gibbon, common *18*
giraffe 23, 30–1, 163; *15*
giraffe, Masai 30
giraffe, reticulated 30
glow-worm 147, 148
goat 13, 31
goby 125
Goliath beetle 138
gorilla 17–18, 19; *18*
gorilla, mountain 19
grasshopper 139, 148, 173
great skua 74
grebe, western *82*
greenfinch 68
green monkey 16
green toad 89
grey tree-frog 90

grey twig snake 104
grizzly bear 21
grouper 125
grouse, black 70
grouse, sage 69–70
grouse, sharp-tailed 70
guanaco 28
guillemot 74–5
gull 68, 75, 165
gull, herring 74
gull, laughing 71

haddock 116
hagfish 113
hare 56
hare, snowshoe or varying 47
hatpin sea-urchin 116
hawk 67
hawkmoth 7, 151
hawkmoth; elephant *136*
hedgehog 14
heron 66, 75, 165
herring 113, 120; *129*
herring gull 74
Hesperornis 165; *165*
hippopotamus 25, 163; *39*
Hoehnel's chameleon 100
honey badger 52–3
honey guide bird 52
horned puff-adder 103
horse 15, 27, 163
hummingbird, ruby-throated 79, 80
hyena 163
Hyracotherium 163
hyrax 163

ichneumon fly *157*
Ichthyornis 165; *164*
ichthyosaur 171; *170*
Ichthyostega 168; *168*
Iguanadon 169
impala 23
Indian hill mynah 73
Indian rhinoceros 27; *38*
insect 13, 136–152, 172–3
Irish elk *162*

jackal 54
Jackson's chameleon 101
Japanese macaque 48; *57*
Javan rhinoceros 27–8
jay 73
jay, Mexican 72
jelly-fish 116

kangaroo 15, 43–4
kangaroo, red *44*
killer whale 14, 26–7
kiwi 137
Komodo dragon 101; *105*

lace monitor 102
lacewing 173
ladybird 149
lance-headed snake 92
lanternfish 114, 128
lamprey 113, 166; *115*
laughing gull 71
leaf-fish 124–5
lemming, Norway (brown) 46–7
lemur, ring-tailed *40*
leopard *59*
leopard snake 103
lesser Malay chevrotain 31
linnet 68
lion 22–3, 30, 163; *15, 35*
lizard 89, 91–2, 98, 162, 168, 169

lizard, spiny-tailed 91
lizard, viviparous 91
lizard, worm 91
llama 28
loach 113
loggerhead shrike 78, 96
long-eared owl *64*
lumpsucker 113
lungfish 167

macaque, Japanese 48; *57*
macrurid or rat-tailed fish 119
Madeiran storm petrel 76
mako shark 116
mammoth 163; *162*
man *15, 18*
mangrove monitor 102
manta ray 113, 126
mantis, flower *153*
marlin 126
marmot 13
marsh frog 90
marsh turtle 91
marsupial 43–4, 55, 162
marsupial anteater 49
marsupial frog 89, 95
marsupial mouse 49–50
marten 13
Masai giraffe 30
masked booby 76
matamata 96
mealworm 150
Megalocephalus 169
Meganeura 173; *172*
Metoedischia 173; *173*
Mexican cave-fish 118
Mexican jay 72
Mexican pygmy salamander 90
Mexican tetra 118
midwife toad 90
mockingbird 73
mole 14
mole rat 13
mole salamander 93
mongoose 20
monitor 97, 101–2
monitor, Bengal 102
monitor, desert 101
monitor, desert pygmy 1–2
monitor, emerald 101
monitor, lace 102
monitor, mangrove 102
monitor, Nile 101
monitor, short-tailed pygmy 102
monkey 13, 17, 48
monkey, green 16
monotreme 43, 49
mosquito 137, 139; *138, 158–9*
moth 137–8, 147, 151
moth, cinnabar 149
moth, tiger 151
mountain chameleon 101
mountain gorilla 17
mouse 47
mouse deer 31
mouse, marsupial 49–50
mouthbrooder, African *130*
mullet 125
musselpecker 76
mynah 73
mynah, Indian hill 73

natterjack 90
newt 89, 90, 170, 171
newt, European mountain 90
nightingale 10
Nile crocodile 97–8

Nile monitor 101
Norway lemming 46–7; *47*
numbat 49
nymph 139, 140

okapi 30
Ophidorpeton 169; *168*
orang-utan 19; *18*
orchids 8
ostrich 66, 77, 169; *83*
osprey 75
Ornithomimus 171
owl 67
owl, eagle 47
owl, long-eared *64*
ox 31
oxpecker 28
oystercatcher 76

Pacific Ridley turtle *110–11*
panda, giant 21–2
panda, red 22
pangolin 14, 49; *48, 60*
paradise tree-snake 104
paradise widow bird 73, 74
parrot 75
parrotfish 11, 115, 125
peacock 70
pelican 75, 165
penguin 67, 75
penguin, Adélie 74, 75
penguin, emperor 71; *71*
Phororhacos 165; *164*
pig 17, 100, 101
pig-deer 101
pike *131*
pipefish 124
pit-viper 92, 103
placental 43
plankton 26, 113
platypus, duckbilled 43; *43*
Plesiosaur 169
poison arrow frog *88*
polar bear 20–21, 163;
   (distribution) *20, 21, 36*
porcupine 55
porcupine fish *112*
porpoise *15*
Portuguese man-of-war 115
prairie chicken 70
prairie dog 50
prairie marmot 50
pronghorn *15, 31*
pseudoscorpion *157*
ptarmigan, willow *85*
Pteranodon 171; *170*
Pterodactylus 169
Pteroplax 169
Pterosaur 169
puff adder 103
puffin 75; *86*
puma 41
pupfish 115

rabbit 54, 56
raccoon 22
rat 16, 142
rat, black 56
rat, desert 45
rat, giant 13
rat, mole 13
rat snake 103
ratel 52
rattlesnake 50, 92
rattlesnake, tropical 103
ray 113, 167
ray, manta 114

ray, Mediterranean electric 120
razorbill 75
red-footed booby 76
red fox 53, 54
red panda 22
red-vented weaver 71
reindeer 31, 163
remora 126; *131*
reptile 13, 91–4, 99, 168–9, 170
reticulated giraffe 30
rhinoceros 27–8, 125, 162, 163
rhinoceros, black 27, 28
rhinoceros, Indian 27; *38*
rhinoceros, Javan 27
rhinoceros, Sumatran 27
rhinoceros, white 27
rhinoceros, woolly 163
Rhyssa 157
ringed seal 20
rivulin 114
robin 67
rodent 13, 16, 41, 44, 50, 55–6,
   71
roe deer 44
ruby-throated hummingbird 79,
   80
ruff 70
ruminant 31; (stomach) *31*

sabre-toothed dragon fish 114
sabre-toothed tiger 14
sage grouse 69–70
salamander 89, 90, 93, 170, 171
salamander, alpine 80
salamander, giant 90
salamander, Mexican pygmy 90
salamander, mole 93
salamander, Texan blind 94
salamander, tiger 93
salmon, Atlantic 117; (migration)
   117, 134–5
sand-sel 116
sand martin 79
sardine 113
scorpion 137
sea-cow 9, 163
sea-lizard 91
sea-otter 25
sea-snail 113
sea-urchin 91, 92, 124
sea-urchin, hatpin 116
seal 14, 44
seal, ringed 20
Seychelles giant tortoise 96
Seymouria 169; *169*
shark 113, 114, 126, 167; *114*
shark, basking 114
shark, blue 116
shark, mako 116
shark, whale 113
shark, white *132–3*
sharp-tailed grouse 70
sheep 13, 31
shrew 13, 14, 162
shrew, Etruscan 13
short-tailed pygmy monitor 102
shrike 78
shrike, loggerhead 78
shrimpfish 116
siamang, dwarf *18*
sidewinder 91

silkmoth 148
silverfish *172*
skimmer 75
skink, common 91
skink, tree 104

skua, great 74
slow-worm 91, 92
snail, sea 113
snake 89–92, 102–4, 168, 169
snake, Aesculapian 103
snake, American green whip 104
snake, Asian green whip 104
snake-bird 87
snake, chicken 103
snake-eel 124
snake, grey twig 104
snake, leopard 103
snake, paradise tree 104
snake, rat 103
snake, spotted rat 103
snow geese *84*
snowshoe hare 47
sociable weaver 71
spade fish 124
spadefoot toad 94
spear-nosed bat 51–2
sperm whale 25–6
spider 137, 147
spiny anteater 49; *49*
spiny-tailed lizard 91
spoonbill 75
spotted rat-snake 103
sprat 116
springbok 15
springtail *173*
squid, giant 25, 26
squirrel 13
stag 16
stareater 114, 128
starling 73
Stegosaurus 171
Stenodictya 173; *173*
stingray 121
stonefish 121
stork 75
storm petrel 76
storm petrel, Elliot's 76
storm petrel, Galapagos 76
storm petrel, Madeiran 76
stromateid fish 115
Sumatran rhinoceros 27–8
suckerfish 126
surgeon fish 122
swallow 66, 67, 71, 79
swallowtail *139*
swift 66, 67
swift, spine-tailed *67*
swiftlet 52
swordfish 126

tadpole 89, 94
tapir 27
termite 48, 139; *154–5*
tern 68–9, 75
tetra, Mexican 118
Texan blind salamander 94
Thomson's gazelle 23
tickbird 125
tiger 22–3
tiger-moth 151
tiger salamander 93
timberwolf, Canadian 54
toad 89–90, 94, 170, 171
toad, bowl-backed tree 89, 95
toad, common 90
toad, green 89
toad, midwife 90
toad, spadefoot 94
toad, Surinam 95
toad, Surinam pygmy 95
toadfish 119

tokee *101*
tooth-carp 115
toothed whale 13, 25–6
tortoise 43, 91, 92, 96
tortoise, Galapagos giant 96
tortoise, Seychelles giant 96
Triceratops 171; *170*
trigger fish 116
tripod fish 114
tropical rattlesnake 103
tuatura 92, 98–100; *98*
tuna 114, 116, 126
turbot 120
turkeyfish 120
turtle 91, 92, 96, 169
turtle, alligator snapping 96
turtle, fringed 96
turtle, marsh 91
turtle, Pacific Ridley *110–11*
turtle, snake-necked 96
Tyrannosaurus 171; *171*

Uintatherium 163; *162*
unigulates 27, 28, 31, 32
Urodela 90

vampire bat 13, 51
varying hare 47
vicuna 24
viper 92; *102*
viper, Gaboon 92, 102–3
viper, pit 92, 103
viper, rhinoceros horned 103
viperfish 127–8
Virginian deer *61*
vizcacha 71
vulture 77
vulture, Egyptian 77; *77*

wagtail 79
walleye pollack 116
warbler 66, 80; (migration) *79*
warbler, willow 10
warblerfinch 66
wasp 141–2, 146, 149
waxbill 73
weasel 20, 100
weaver ant 146
weaver bird 71–2; *72*
weaver, red-vented 71
weaver, sociable 71
weever fish 121
western grebe *82*
whale 13, 14, 25–7, 126, 163
whale, baleen 13
whale, blue 13; *14–15*
whale, humpback *37*
whale, killer 14, 26–7
whale shark 113
whale, sperm 15–16
whale, toothed 13, 15–17
white-tailed deer *61*
whiting 116
wildebeeste 23
willow ptarmigan *85*
wolf 41, 43, 53, 54, 163; *23*
wolverine 41, 47, 53
woodhewer 71
woodlouse 137
woodpecker-finch 66
woolly rhinoceros 163
worm lizard 91
wrasse 11, 123, 125–6
wrasse, cleaner 11, 123
wrasse, West Indian 123

zebra 23, 27